基于步行网络可达分析的公平性评价与优化模型

徐孟远 著

武汉大学出版社

图书在版编目(CIP)数据

基于步行网络可达分析的公平性评价与优化模型/徐孟远著.—武汉：武汉大学出版社,2016.12
　ISBN 978-7-307-18813-6

Ⅰ.基… Ⅱ.徐… Ⅲ.城市公用设施—城市规划—研究　Ⅳ.TU99

中国版本图书馆 CIP 数据核字(2016)第 275207 号

责任编辑：黄金涛　　责任校对：汪欣怡　　版式设计：马　佳

出版发行：**武汉大学出版社**　(430072　武昌　珞珈山)
　　　　　(电子邮件：cbs22@whu.edu.cn　网址：www.wdp.com.cn)
印刷：虎彩印艺股份有限公司
开本：720×1000　1/16　　印张：14.5　　字数：378 千字　　插页：1
版次：2016 年 12 月第 1 版　　2016 年 12 月第 1 次印刷
ISBN 978-7-307-18813-6　　定价：55.00 元

版权所有，不得翻印；凡购我社的图书，如有质量问题，请与当地图书销售部门联系调换。

CONTENTS

CHAPTER ONE　INTRODUCTION ··· 1
  1.1　Background ·· 1
  1.2　Research Question ·· 5
  1.3　Research Goals ·· 5
  1.4　Contents of the Document ·· 6

CHAPTER TWO　LITERATURE REVIEW ·· 8
  2.1　Walkability and Pedestrian Environment ································· 8
  2.2　Life-Needs Service Facilities ··· 13
  2.3　Nature of the Topic of Services Distribution ··························· 20
  2.4　Notion of Equity ··· 21
  2.5　Notion of Accessibility ··· 24
  2.6　Measures of Accessibility ·· 26
  2.7　Defining Target Groups ·· 37
  2.8　Trade-Off between Equity and Efficiency ······························· 38
  2.9　Geographic Information System (GIS) Platform ······················· 40

CHAPTER THREE　METHODOLOGY ·· 42
  3.1　Theoretical Framework ··· 42
  3.2　Data Preparation Procedure ··· 43
  3.3　Data Analysis Procedure ··· 56

CHAPTER FOUR　RESULTS ··· 70
  4.1　Census Block Centroids Locations ·· 70
  4.2　Route Calculation ··· 73
  4.3　Intersected Sidewalk from the Routes ··································· 84
  4.4　Double-Checking Missing Sidewalk Segments on Both
      Sides of the Roads ··· 98

4.5　Efficiency and Equity Importance Measure Calculation ··················· 100
4.6　Prioritization Results of Various Scenarios ························· 100

CHAPTER FIVE　DISCUSSION AND CONCLUSIONS ·················· 117
5.1　Discussion ·· 117
5.2　Limitations ·· 120
5.3　Future Study ·· 121

BIBLIOGRAPHY ·· 122

APPENDIX ·· 132
　Appendix A: List of all Daily Food Providers within the Study Area ············ 132
　Appendix B: List of all Social Gathering Facilities within the Study Area ······ 134
　Appendix C: List of all Sports and Recreation Facilities within the
　　　　　　　Study Area ·· 156
　Appendix D: List of all Healthcare Facilities within the Study Area ············ 162
　Appendix E: Demographic Data of Focus Census Blocks ························ 196
　Appendix F: Missing Sidewalk Segments with Equity Importance ··············· 212
　Appendix G: Highest Priority Segments When $r=1$ ··························· 215
　Appendix H: Highest Priority Segments When $r=2$ ··························· 217
　Appendix I: Highest Priority Segments When $r=0.5$ ························· 219
　Appendix J: Highest Priority Segments When $r=0$ ··························· 221

LIST OF TABLES

1. Table 1: Focus census block groups identified by vehicle ownership rate. ············ 73

LIST OF FIGURES

1. Figure 1: Position of various accessibility measures in the resource-midfare-welfare sequence. 35
2. Figure 2: Trade-off between equity and efficiency. 40
3. Figure 3: Spokane pedestrian network creation. 46
4. Figure 4: Polygons divided by road network in the Spokane metropolitan area. ... 47
5. Figure 5: Missing sidewalks data process. 49
6. Figure 6: Entrance identification by satellite map and street view image for Manito Park with Google Street View. 51
7. Figure 7: Edited point feature class of park entrances in ArcMap Editor. 52
8. Figure 8: Life-Needs Service Facilities locations recorded as point feature classes. 53
9. Figure 9: Vehicle ownership rate calculation at the level of census group block. 54
10. Figure 10: Focus census blocks selected from census block groups with a low vehicle ownership rate. 55
11. Figure 11: Overall data analysis flowchart. 58
12. Figure 12: Route from census block 1036 to the closest daily food provider. 61
13. Figure 13: Route from census block 1036 to the closest social gathering facility. 62
14. Figure 14: No healthcare facility or park entrance from census block 1036 within walking distance. 63
15. Figure 15: Double-checking missing sidewalk segments on both sides of the roads. 67
16. Figure 16: Prioritization of missing sidewalk segments. 68
17. Figure 17: Census block centroids within Spokane PTBA. 71
18. Figure 18: Focus census block centroids distribution. 72
19. Figure 19: Routes from all census block centroids to the closest daily food provider within walking distance. 74

LIST OF FIGURES

20. Figure 20: Routes from all census block centroids to the closest social gathering place within walking distance. 75
21. Figure 21: Routes from all census block centroids to the closest sports and recreation facility within walking distance. 76
22. Figure 22: Routes from all census block centroids to the closest healthcare facility within walking distance. 77
23. Figure 23: Routes from all census block centroids to the closest park entrance within walking distance. 78
24. Figure 24: Routes from focus census block centroids to the closest daily food provider within walking distance. 79
25. Figure 25: Routes from focus census block centroids to the closest social gathering place within walking distance. 80
26. Figure 26: Routes from focus census block centroids to the closest sports and recreation facility within walking distance. 81
27. Figure 27: Routes from focus census block centroids to the closest healthcare facility within walking distance. 82
28. Figure 28: Routes from focus census block centroids to the closest park entrance within walking distance. 83
29. Figure 29: Census blocks that have access to all five categories of Life-Needs Service Facility. 85
30. Figure 30: Census blocks that have no access to any Life-Needs Service Facility. 86
31. Figure 31: Missing sidewalk segments selected from routes linking all census block centroids to the closest daily food provider within walking distance. 87
32. Figure 32: Missing sidewalk segments selected from routes linking all census block centroids to the closest social gathering place within walking distance. 88
33. Figure 33: Missing sidewalk segments selected from routes linking all census block centroids to the closest sport and recreation facility within walking distance. 89
34. Figure 34: Missing sidewalk segments selected from routes linking all census block centroids to the closest healthcare facility within walking distance. 90
35. Figure 35: Missing sidewalk segments selected from routes linking all census block centroids to the closest park entrance within walking distance. 91

LIST OF FIGURES

36. Figure 36: Missing sidewalk segments selected from routes linking focus census block centroids to the closest daily food provider within walking distance. 92
37. Figure 37: Missing sidewalk segments selected from routes linking focus census block centroids to the closest gathering place within walking distance. 93
38. Figure 38: Missing sidewalk segments selected from routes linking focus census block centroids to the closest sports and recreation facility within walking distance. 94
39. Figure 39: Missing sidewalk segments selected from routes linking focus census block centroids to the closest healthcare facility within walking distance. 95
40. Figure 40: Missing sidewalk segments selected from routes linking focus census block centroids to the closest park entrance within walking distance. 96
41. Figure 41: Missing sidewalk segments eliminated from selection. 99
42. Figure 42: Result of sidewalk segment improvement prioritization (when $r=1$). 101
43. Figure 43: Result of sidewalk segment improvement prioritization (when $r=2$). 102
44. Figure 44: Result of sidewalk segment improvement prioritization (when $r=0.5$). 103
45. Figure 45: Result of sidewalk segment improvement prioritization (when $r=0$). 104
46. Figure 46: Result of prioritization calculation based on equity importance measure. 106
47. Figure 47: Result of prioritization calculation in Area #1 (when $r=1$). 107
48. Figure 48: Result of prioritization calculation in Area #1 (when $r=2$). 108
49. Figure 49: Result of prioritization calculation in Area #1 (when $r=0.5$). 109
50. Figure 50: Result of prioritization calculation in Area #1 (when $r=0$). 110
51. Figure 51: Result of prioritization calculation in Area #1 based on equity importance measure. 111

LIST OF FIGURES

52. Figure 52: Result of prioritization calculation in Area #2 (when $r=1$). ·· 112
53. Figure 53: Result of prioritization calculation in Area #2 (when $r=2$). ·· 113
54. Figure 54: Result of prioritization calculation in Area #2 (when $r=0.5$). ·· 114
55. Figure 55: Result of prioritization calculation in Area #2 (when $r=0$). ·· 115

CHAPTER ONE   INTRODUCTION

This chapter presents an introduction to this research and provides a brief background, the research context, the existing problem and justifications, research questions, the research goals, and definitions of terms.

1.1 Background

Walking is the most basic and fundamental form of transportation (Lee & Moudon, 2006; Litman, 2003). The terms "walkability" or "walkable" have been widely used in research papers related to built environments and walking behavior. However, conventional planning tends to pay less attention to walking activity compared to motorized travel.

Conventional planning considers walking as a minor mode of travel because it represents only one fiftieth the amount of vehicle travel in terms of person-miles (Litman, 2003). Therefore, the quality of walking environments has received little support in conventional planning (Moudon et al., 2008). However, more recently, the concepts of walkable communities and walkability have been receiving increasing research interest and attention with the growing body of findings that walkability is related to public welfare (Moudon et al., 2008; Ewing et al., 2006). Researchers from the healthcare field have been contributing a flourish of literature on the impacts of built environments on people's physical activity, which strongly influence people's health (Eyler et al., 2003; Haskell et al., 2007). Walking has also been considered by researchers to be vital for increasing social interaction with the local community (Leyden, 2003). In addition, walking itself also serves as a crucial part in the overall transportation system, especially for mobility-disadvantaged groups who lack basic mobility in an automobile-dependent community (Litman, 2003).

Among all the relative parts of the characteristics of a pedestrian environment, the distribution of critical urban services among urban residents has been a major focus in urban planning and policy making worldwide, because access to crucial life-needs services such as grocery stores, gathering places and green spaces are vital to people's welfare and

CHAPTER ONE   INTRODUCTION

well-being (Beaulac et al., 2009; De Vries et al., 2003). The study of measuring access to public goods and services has received much interest for decades.

Considering that the study of accessibility deals with the distribution of certain urban resources among certain groups of people, it is naturally related to the notions of equity as well as efficiency. Both equity and efficiency have been considered critical factors to improve the performance of an urban environment (Talen, 2011; Dietz & Atkinson, 2010; le Grand, 1990). Equity is related to demands and available resources among social groups, and efficiency is how well the infrastructure serves the groups to a maximized degree. However, although equity is widely considered as a vital principle among policy makers and planners (Talen, 1997), the notion of equity remains a very ambiguous term and is usually addressed from a broader view of social analysis (Smith, 1994). There is no common agreement on what an equitable distribution would be due to the complex nature of the issue (Blanchard, 1986).

At the same time, the notion of efficiency is interpreted differently by different researchers, which leads to a variety of measures of efficiency. In other words, despite continuing research interest and numerous geographic studies that have discussed and contributed to the notions of equity and efficiency, researchers may not mean the same thing when they use the concepts due to the fact that transportation equity analysis of accessibility involves multiple dimensions of issues and various approaches to define and measure them (Litman, 2003). Considering that the lack of consensus on a consistent understanding of the notions would hamper further discussion on this issue, it is necessary to firstly clarify and define the notions of equity and efficiency based on the topic at hand in a transportation planning context.

A theoretical foundation is proposed with a definition of the relative notions of walkability and accessibility in the context of this research. The comprehension of this conceptual set not only leads to a framework that provides comprehensive grounds for future research, but also may make progress towards a new sophisticated way of analyzing equity that planners and researchers can use. Furthermore, when it comes to the destinations to which access should be provided, there is a lack of a consensus in existing literature on the framework of categories of critical urban services, which is referred to here as Life-Needs Service Facilities. A well-formed category of the service facilities can be of large assistance for future discussion and model building.

The primary objective of research on measuring the accessibility to urban service facilities based on spatial and demographic patterns is to propose a potential prioritization of improvements to achieve a more efficient or equitable system. Despite continuing research interest in this field, the question of whether or not these social services are equally distributed has "largely remained unsolved" (Talen, 1996). The performance measures of the efficiency of a system are developed in very different ways based on various focuses and scales of the research. So, there is a great need for a sophisticated model with quantitative accessibility measures to study walking activity and evaluate the pedestrian environment. In addition, in the context of limited resources and budgets for delivering future plans, it is vital to consider how to improve the way that services should be delivered. In order to achieve maximized benefits, approaches are required to identity the importance of each specific segment of the infrastructure in the overall network in order to decide their prioritization for fixing, and a number of previous studies have proposed different approaches. As a result, it is vital to prioritize future improvements based on their importance by responding to these criteria.

Computer-aided design technology has been widely used in the process of design since the 1980s. However, the roles of most current computer-aided design tools (such as AutoCAD and Photoshop) are generally limited to drafting and documenting. Similar to traditional drafting boards and manual drawings, these traditional design tools still act as "representational tools," which are facilitated to visually represent either existing conditions or proposed designs while not actively participating in the design thought process or the decision-making process (Kotnik, 2010). It is considered that we are currently at the digital threshold of the evolution of computer-aided design tools from representational tools to the next stage of "simulation tools." Facilitated by quantitative spatial analysis tools, simulation tools are able to provide in-depth information related to design goals and thus optimize the decision-making process. The shift of paradigm from representational tools to the next stage represents the trend of computer-aided tools evolving from representative tools of describing "what it is" to stimulation tools that are able to describe "what can be" based upon various proposals. In other words, tools in the next stage would support more active design decisions, helping designers to find the optimal solution by providing real-time feedback to proposed design scenarios.

With powerful abilities to store, manage, analyze and present spatial data, GIS provides great convenience for planners to create models to simulate a built environment. GIS enables the operation of spatial measures of the built environment that provide spatial linkages that integrate demographic data, spatial patterns, and other relative environmental factors with the outstanding capacity of enabling visual representation. These advantages help the decision-making process for urban planners and make the GIS an ideal platform of analysis. The concepts of GeoDesign and Geovisualization are adopted to reflect

CHAPTER ONE   INTRODUCTION

this shift. First proposed by Dangermond (2008), the concept of GeoDesign was to integrate geographic analysis into the design decision making process. A GeoDesign platform is able to combine a variety of database layers describing social and physical factors related to the design project. This platform provides a framework that allows a GeoDesign model to draw different design scenarios in digital format, which not only represents the design product, but more importantly, provides analysis tools that assist designers to explore the optimal design scenario based on certain criteria. For the pedestrian accessibility analysis in this study, socio-demographic factors are considered as significant components in assessment and prioritization process. The GeoDesign concept provides guidance to integrate socio-demographic layers into a spatial analysis model to create assessment and prioritization tools, which can be used to assist urban planners and policy makers to explore optimal design scenarios. On the other hand, the concept of Geovisualization represents the idea that the model would take advantage of the ability to visually present changes in real time and allow for a more interactive process, which would better facilitate the planning process among various stake holders (Talen, 2011). The concept of Geovisualization is also reflected in this study. This analysis model provides a customizable prioritization criteria and the model is able to visualize different scenarios based on different settings.

   Recently, some analysis tools based on the GIS database have been developed by researchers. However, existing accessibility equity measures suffer from some disadvantages. First, transportation equity analysis is highly affected by the modes on which the research focuses. The majority of existing research favors motorized accessibility, and relatively little attention has been paid to non-motorized or pedestrian accessibility (Litman, 2003). The notion and measures of pedestrian accessibility still have not been thoroughly discussed, and this paper fills the gap in the literature. Furthermore, even in recent physical activity research focusing on the pedestrian environment, street networks tended to be used for connectivity analysis of the built environment, whereas a true pedestrian network might be very different from a street network in regard to connectivity calculation (Chin et al., 2008).

   Another significant deficiency is that previous measures tended to divide individual socio-economic and socio-demographic status from their geographical context (Weber & Kwan, 2003). Considering that the level of accessibility is largely affected by the social characteristics of households rather than just the built environment, it is necessary to relate accessibility to social inequity and social stratification (Hanson & Pratt, 1988). So, there is a need to further examine the association between accessibility and social characteristics of the population groups.

This significance of this study is its thorough discussion of the ambiguous notions of equity and accessibility from a transportation planning standpoint. Addressing the needs for filling gaps in existing studies, it synthesizes studies from various fields, including urban planning and sociology, and establishes a theoretical framework for future studies. This book discusses the significance and development of a methodology for creating a GIS-based accessibility measurement model in the Spokane PTBA (Public Transit Benefit Area) for mobility-disadvantaged groups based on information from previous studies. Significant improvement in the methodology and tool development can be drawn from this research. An integrated GIS tool was designed to provide broad spectrum analysis and real-time feedback for equitable as well as efficient urban development. The model not only helps visually illustrate potential accessibility inequities to life-need service locations across the whole study area, but also shows potential changes in terms of the level of accessibility based on the improvement of pedestrian facilities or newly added service locations. It is also a design-decision support tool that helps planners identify the priority of improvement and find optimal scenarios based on principles of efficiency and equity.

1.2 Research Question

This research addresses the question of how to develop a GIS-based model that can measure pedestrian accessibility and prioritize potential improvement based on the criteria of equity and efficiency. In order to answer this question, there are a few aspects that need to be explored. First, it is necessary to review the measures of accessibility (or the distribution of urban services among citizens) in the context of transport and urban planning. Second, it is important to understand how to establish the accessibility model of a pedestrian environment on a GIS platform. Third, it is necessary to clarify the notions of equity and efficiency in the context of urban planning and to propose how to interpret these two notions in the assessment and prioritization model. These topics are discussed in Chapter 2 in detail.

1.3 Research Goals

The aims of this research are to achieve the following goals:

(1) *Contribute to the literature on the notions of equity and efficiency in terms of accessibility to critical urban services*. Both equity and efficiency have been ambiguous terms

in previous studies due to various research focuses and opinions. So, the primary object is to review the relevant literature and then to thoroughly discuss and understand the notions of equity and efficiency in terms of accessibility analysis and related social factors in the context of transportation planning.

(2) *Creation of a sophisticated GIS-based model of the pedestrian environment in the study area.* Due to its powerful functions in data management, analysis, and representation, GIS is an ideal platform for the model of pedestrian environment. The previous models of pedestrian environment were created in different ways and may have certain disadvantages. For example, most current accessibility analysis researches employ street network based instead of true pedestrian network, which could be misleading (Chin et al., 2008). Based the present research question, the first objective is to discuss how pedestrian environment can be represented and modeled in the GIS. The second objective is to propose a model with detailed demographic and geographic data on the Census block level as well as true pedestrian network data.

(3) *Proposal of pedestrian accessibility measurement approaches in this GIS-based model so it can evaluate the accessibility patterns in the study area.* Previous research implemented various approaches to measure the accessibility based on various research contexts, so it is important to conduct a comprehensive review of these different approaches before proposing an appropriate measure for this specific research topic.

(4) *Developing a methodology to prioritize potential infrastructure improvements in the context of limited resources and budgets based on the trade-off of equity and efficiency.* The notions of both equity and efficiency are ambiguous in previous studies. As a result, it is necessary to clarify these two notions in the present research and thus interpret them into measures in the prioritization model.

1.4 Contents of the Document

This book is comprised of five chapters. The beginning of each chapter contains a short summary of the subject to be covered. The first chapter is the introduction which presents a brief background of the research context, research question, and research goals. The second chapter is the literature review. It presents the theoretical background of the study which provides a review of the research topics, including walkability, pedestrian environment, notion of accessibility, notion of equity and efficiency. The third chapter presents the theoretical framework and methodology of data preparation as well as data analysis procedures that are used in the research. Figure 11 shows the overall data analysis procedures. The fourth chapter presents the results of the research model applied to Spokane Public Transit Benefit Area. The fifth chapter contains the discussion and limitation of the study, and potential future study.

CHAPTER TWO LITERATURE REVIEW

The purpose of the literature review is to provide a thorough overview of the fields of urban service distribution and accessibility analysis in transportation planning literature and answer the first research questions. The specific topics that are covered in this chapter include: walkability and pedestrian environment; urban services categorization; notion and measurements of accessibility; notion of equity and efficiency; trade-off between equity and efficiency; Geographic Information System (GIS) platform.

2.1 Walkability and Pedestrian Environment

Along with increasing research interest in the relationship between built environments and walking behavior in recent years, the terms "walkability" or "walkable" have been widely used in relevant research papers (Moudon et al., 2006). However, only of few of them have provided clear definitions of either term. In a paper submitted during the 2004 USA Transportation Research Board Meeting, Livi and Clifton (2004) also pointed out that none of the papers directly "explained and defined the term." In a paper presented to Land Transport New Zealand, Abley (2005) proposed the definition of **Walkability** as **"the extent to which the built environment is walking friendly."** This definition was also quoted in the Pedestrian Planning and Design Guide (2008) developed by the New Zealand Transport Agency. The level of walkability of a neighborhood is widely considered to relate to the physical characteristics of the built environment, such as density of neighborhood, street connectivity, and distance to service facilities (Ewing et al., 2006).

The rapid development and introduction of the automobile has had a huge impact on shaping city patterns over the last century. The use of private automobiles has brought about huge convenience and overall mobility of people and needed goods. Motorized transportation has become a major catalyst that has stimulated the rapid development of suburbs and the decay of the inner city since the 1940s (Duany et al., 2001). The vast investment in highway infrastructure in the post WWII period further strengthened the dominant status of private automobiles as the preferred mode of transit in the U.S.

(Jacobson, 1985). This phenomenon was referred to as "urban sprawl" in the field of urban planning (Duany et al., 2001). From the perspective of conventional urban and transportation planning practice, walking is just a minor mode of travel and has received little attention and support (Litman, 2003).

Along with the separation of land uses and automobile-oriented transportation planning due to suburban sprawl, inner city neighborhoods have become predominantly occupied by lower social status groups (Algert et al., 2006). By the 1970s, commercial service providers (for example, stores) also moved out of the inner city to newly developed suburbs that were closer to their main customers (Pothukuchi, 2005). As a result of shifts in transportation modes and land-use patterns, the level of spatial accessibility to various service providers has been dramatically changed in large North-American metropolitan areas (Larsen, 2008). However, even today there are still a large portion (9.2% in 2012) of overall households that do not own a private car (U.S. Census Bureau, 2012) and many mobility impaired individuals, such as elderly and children, may have limited access to a car. As a result, it is widely believed that the disparities in access to essential services between advantaged and disadvantaged communities (usually low-income inner-city neighborhoods) have been amplified because of differences in vehicle ownership rates (Bader et al., 2010; Ford and Dzewaltowski, 2008). Under these circumstances, walking as an alternative travel mode has played an important role in accessing needed services for these disadvantaged groups and has acted as a significant component in the overall transportation system (Litman, 2002; Handy & Clifton, 2001; Iacono et al., 2010; Ploeg, et al., 2009).

On the other hand, studies on walkability and walkable neighborhoods have been extending beyond mere transportation concerns (Moudon et al., 2006). Researchers from the healthcare field have found ample evidence suggesting that the structure of a built environment has impacts on people's physical activity and thus strongly influences people's health (Cao & Mokhtarian, 2012; Moudon et al., 2006). It is widely argued that residents living in neighborhoods with high accessibility tend to walk more than those living in neighborhoods with low accessibility (Cervero & Duncan, 2003; Joh et al., 2008; Chatman, 2009). In addition, supporters of New Urbanism advocate walkable neighborhoods as a way to enhance sociability among citizens, which benefits spiritual health of people in the community (Moudon et al., 2006; Duany et al., 2001). These mounting empirical findings have motivated local governments to facilitate walking-friendly land-use policy and design guidelines to promote the non-automobile motorized modes and reduce the use of vehicles for environmental reasons (Cao & Mokhtarian, 2012). For example, California's Senate Bill 375 —"Redesigning Communities to Reduce Greenhouse Gas" — was proposed to support implementations of sustainable community strategies and reduce greenhouse gas emissions through land use and transportation strategies. Combining all these arguments, it is widely believed that walking itself is an important activity that affects people's physical and mental health (Litman, 2003; Moudon et al., 2006).

Due to its importance, the concept of walkability and walkable neighborhoods has recently received increasing research interest for both transportation and health purposes. The first conceptual model of neighborhoods in urban planning history is not very old. In 1929, Clarence Perry proposed the model of a "Neighborhood Unit." The structure of this model was defined based on the distance that residents were able to walk safely and easily from their homes to schools, community centers, and churches within each neighborhood unit. Perry's conceptual model and ideas were realized in neighborhoods such as Radburn and hugely impacted the theory and practices of neighborhood planning afterwards. Perry proposed that the walking distance from home to community center should not be more than 1/4 mile (400m) in a neighborhood unit.

More recently, the most influential planning theory and practices related to walkable communities came from the "New Urbanism" movement, which was first proposed by Andres Duany and Elizabeth Plater-Zyberk in the 1990s. In their notable book, *Suburban Nation: The Rise of Sprawl and the Decline of the American Dream*, Duany (2001) and his colleagues strongly criticized the urban sprawl in cities guided by conventional planning and considered typical suburbs as alienating and a "waste of land." They accused automobile-oriented design and the separation of land use of hampering the accessibility the neighborhoods and damaging the sense of community. Instead, they proposed that neighborhoods be designed in a high-density and "pedestrian-friendly" way in which residents live within walking distances of essential urban services.

To further understand the neighborhood walking environment, the development of measures is essential (Brownson et al., 2009). A number of studies have addressed the issue of evaluating pedestrian environments, and several approaches have been formulated to describe and simulate the characteristics of built environments (some of which include crime rate/security, which is not completely within the domain of built environment, but the majority variables in these measures are related to the built environment). There is no evaluation system that is accepted by all so far due to various research foci and theoretical foundations. Based on the data collected to describe the built environment, these measures can be divided into two categories: perceived measures and objective GIS-based measures. Here is a brief review of several typical measures proposed or employed in previous studies.

In perceived measures, the data used is based on the perception by people of the built environment and is usually collected through self-reported approaches, such as surveys and interviews (Kelley et al., 2010; Cerin, 2009), or audit tools by trained raters (Pikora et al., 2002; Milington et al., 2009). Kelly (2010) and his colleagues developed three survey tools: a stated preference survey, and an on-the-street survey and a mobile survey. The stated preference survey tool was designed to collect the responses from participants in regard to the importance or unimportance of 47 identified attributes on a five-point Likert scale. The on-the-street survey was in a questionnaire form that required the respondents to describe their views and feelings regarding walking on the street. The mobile survey recruited participants by an advertisement in a local paper to take a walk. All participants were unfamiliar with the route they took. During the walk, digitally recorded semi-structured interviews about the walking experience were conducted in order to generate a rich and detailed dataset. Cerin (2009) and his colleagues also developed a perceived measurement tool called The Neighborhood Environment Walkability Scale (NEWS), which includes 38 items considered as important characteristics that influence walking activities. One typical approach based on audit tools was developed by Pikora et al. (2002), who proposed an audit instrument to measure an objective physical environment for physical activity. Their audit tool focused on factors of design (e.g. path surface quality and crossing opportunities), location (e.g. connectivity and potential for vehicular conflict), and user (e.g. pedestrian volume and personal security). Based on Pikora's research, Milington et al. (2009) developed an objective measure tool focusing on urban and suburban contexts called the Scottish Walkability Assessment Tool (SWAT). 112 various items (different destinations/factors) were identified and organized into categories, and 17 items had adequate variability.

In contrast, GIS-based measures refer to "measures of the built environment derived primarily from existing data sources that have some spatial reference, which provides great convenience for characterizing the built environment" (Brownson et al., 2009). These measures were developed based on the powerful abilities of the GIS platform to store, manage, assemble, analyze and present the spatial data from multiple spatial geodatabases (Geurs et al., 2012).

For example, Leslie (2007) and his colleagues proposed a "walking index" instrument-development process based on GIS. The items taken into consideration included (1) measurement of dwelling density; (2) measurement of connectivity; (3) measurement of land use attributes; and (4) measurement of net retail area. Hoehner

(2005) and his colleagues examined the difference between perceptions and objectively measured environmental factors by measuring the association of walking activities and the physical environment through both self-report (perceived) and environmental audit (objective) methods in four urban settings.

One limitation of perceived environment measures is the lack of validity (Brownson et al., 2009). Only a few studies have comprehensively addressed evaluating validity for the measures (Humpel et al., 2002; Pikora et al., 2003; Hoehner et al., 2006). In contrast, the majority of studies to date have used GIS to measure accessibility to services in order to examine spatial inequalities in health care delivery, food environment, and other facilities (Higgs, 2005). One challenge for GIS-based measures is that GIS is a data-hungry analysis model, which means that it requires that all relative datasets be available and manipulated into simplified forms to achieve accurate results (Yigantcanlar, 2007). Developed by ESRI, ArcGIS Network Analyst is a set of extension tools of ArcGIS software, which has become a widely used tool for transportation accessibility analysis by many researchers. It provides sophisticated network-based spatial analysis on the platform of GIS. Using ArcGIS Network Analyst, researchers are able to import and modify data describing the features of a network. It enables users to find the potential routes to destinations or calculate service areas based on certain impedances.

Another main factor is pedestrian network quality, which includes the presence and quality of sidewalks and other necessary facilities (ramps, etc.). Well-maintained sidewalks and facilities are considered positively related to the overall neighborhood walkability level (Cerin et al., 2009; Saelens & Handy, 2008). Milington et al. (2007) also included the presence of path obstacles as one of the potential measures in his research. However, this element proved not to be statistically significant for the overall walkability level in his study.

The previous research that facilitates GIS also suffers from several deficiencies. First, studies based on the GIS platform have proposed many different measures of accessibility, but most of them tended to focus on motorized transport modes (Achuthan et al., 2010; Vandenbulcke et al., 2009). Few of them adequately assess accessibility based on different travel modes (Handy & Clifton, 2001). In addition, ADA accessibility analysis is particularly scarce (Church and Mason, 2003). The most basic measures analyze the demand for certain kinds of services in a defined area, which are commonly Census tracts (Knapp & Hardwick, 2000; Susi & Mascarenhas, 2002).

One significant limitation of these measures is that they assume all households in these areas share equal access to the facilities (Salze et al., 2011), and they do not count the effect of distance in the evaluation of access, which could even be questionable in non-automobile travel modes (Higgs, 2005). To overcome this limitation, a number of recent studies proposed alternative measures which calculate the travel time or distance in GIS to assess the spatial accessibility of the neighborhood (Gutiérrez et al., 2010; Comber et al., 2008). Chin (2008) and his colleagues proposed another concern in access analysis based on GIS in that most research employs street-network-based accessibility analysis, but a true pedestrian network may be very different in connectivity from a street network, meaning that "it is critical that future studies incorporate pedestrian networks into their analyses." For mobility-impaired individuals, the difference between a street network and ADA sidewalk network may be even larger. As a result, the use of a street network instead of a pedestrian network is questionable.

Most pedestrian accessibility-related studies applied the approximate measure of a walking distance, which is 400 to 800 meters (approximately 0.25 to 0.5 miles) in the impedance of distance or 10 to 15 minutes in the impedance of time (Bader et al., 2010). In Larsen and Gilliland's (2009) study in London, Ontario, they considered destination within a 500-meter (0.3-mile) walking distance or a 10-minute (3-kilometer or 1.86-mile) bus ride as accessible. Based on this previous research, a distance of 1/4 mile (400m) was applied as the impedance in the present study.

2.2 Life-Needs Service Facilities

As discussed in the chapter on walkability and pedestrian environments, researchers are interested in an individual's ability to reach desired goods, services, activities and destinations, or "opportunities," as well as how these services are distributed among groups. Litman (2003) described these places as "goods, services and activities that a community considers to have high social value." In a report of the New Zealand Association of Impact Assessment, Baddon (1999) illustrated these services as "a system of social services, networks and facilities that support people and communities." Casey (2005) used the term "social and community infrastructure and services" and proposed that they responded to the "needs of communities." She listed a variety of services and divided them into two categories: "hard" infrastructure and "soft" infrastructure. The former involves the provision of basic utilities that shape the framework in which fundamental activities of a community can be carried out, including water, food, gas, transportation systems, etc. Although also relating to the fulfillment of the needs of communities, the latter focuses on enhancing both the individual's and the community's

CHAPTER TWO LITERATURE REVIEW

well-being by providing a range of equitable community services rather than providing physical assets.

These descriptions from previous studies help depict the broad spectrum of these crucial urban services. However, there is a lack of emphasis and supporting evidence defining what these facilities should be (Manderscheid, 2012). The question of what urban services should be socially important in the study of pedestrian environment has still largely remained unanswered (van Bergeijk et al., 2008). Additionally, there is no literature that has proposed a comprehensive category of urban services. Even for studies that discuss those critical urban service facilities, the categorizations have also been inconsistent. Commonly discussed service facilities in previous research include groceries, social welfare systems, healthcare systems such as hospitals and other medical professionals, leisure and pleasure facilities such as parks and sports centers, and commercial infrastructure such as cinemas and cafes. For example, Van Bergeijk (2008) and his colleagues proposed that groceries, restaurants or cafes, community centers, and libraries were the significant neighborhood facilities. In his research on community infrastructure and services, Carey (2005) listed a variety of important urban services, including healthcare facilities, education, sport and recreation, community development, and emergency services.

To better define and describe opportunities for further study, the notion of **Life-Needs Service Facilities** is proposed, which is defined as a system of physical facilities/services that support the essential needs of individuals, families and communities, and thus enhance the overall quality of life. **Life-Needs Service Deserts** are defined as areas with relatively poor access to Life-Needs Service Facilities, which is considered as an extreme condition of inequity in the distribution of these services.

Due to a lack of precedence from previous academic studies, a categorization of Life-Needs Service facilities is proposed from comprehensive synthesis. The Life-Needs Service Facilities are composed of several categories:

- Daily food providers: grocery stores and supermarkets
- Social gathering places: libraries, community centers, churches
- Sports and recreation facilities: fitness centers/gymnasiums, sports clubs, golf courses, bowling centers, swimming pools, recreation center, and other entertainment and recreation facilities
- Healthcare faculties: hospitals, clinics, and other practitioners that provide practice of medicine
- Parks

2.2 Life-Needs Service Facilities

It is necessary to mention education facilities and schools. Although mentioned in previous research, education facilities and schools are critical for a specific group of citizens (children and teenagers) rather than all citizens. As a result, education facilities and schools are not considered as a general category. The importance of providing access to Life-Needs Service Facilities is illustrated based on different categories.

2.2.1 Daily Food Providers

The daily food providers include supermarkets and local grocery stores, where a wide variety of healthy food can be found at affordable prices (Bader et al., 2010; Walker et al., 2010; Larsen & Gilliland, 2008). The term "food desert" originated in Scotland in the early 1990s and was widely used in previous studies that focused on neighborhood food environments (Cummins, 2007). The concept of "food desert" is defined as neighborhoods that have limited access to affordable and healthy diets (Beaulac et al., 2009; Cummins & Macintyre, 2002). The research interest on food environments was initially motivated by the increasing prevalence of obesity and diabetes (Shaw, 2006; Moore et al., 2008). In the Report of Access to Affordable and Nutritious Food in 2009, the United States Department of Agriculture proposed that these kinds of diet-related diseases had become a major public health problem. This topic has received vast research interest in recent years. Obesity and diet-related diseases are linked with poor diet, and limited access to nutritious foods is one of the more important factors. In the U.S., studies on food access revealed that approximately 6 percent of households cannot always have the food they need due to problems related to poor access (U.S. Department of Agriculture, 2009).

There has been increasing interest in the built environment and availability of healthy food within communities (Walker et al., 2010). Many recent studies have examined the association between food deserts and health outcomes and provide positive correlation between the two consistently (Shaw, 2006; Moore et al., 2008; Lopez, 2007; Powell et al., 2007; Gianget al., 2008; Morland & Evenson, 2009; Michini & Wimbley, 2010). The researchers point out that access to supermarkets or large grocery stores is positively associated with a healthier diet and ultimately diet-related diseases (Bader et al., 2010), while the association with proximity to convenience stores is less statistically positive (Bader et al., 2010; Walker et al., 2010). Indeed, access to convenience stores is associated with a poorer diet and poorer weight status (Morland, et al., 2006; Morland & Evenson, 2009; Grafova, 2008). One key reason could be that convenience stores usually do not provide all the foods needed for a healthy diet, or they provide the foods at higher price (United States Department of Agriculture, 2009). The larger supermarkets are able to offer food in various brands and sizes and thereby keep the price lower, so they have advantages compared to convenience stores and smaller grocery stores (Bader et al., 2010).

For example, Powell (2007) and colleagues examined the association between excess weight and BMI (Body Mass Index) with the availability of grocery stores. The researchers collected data from multiple sources such as the 2000 U.S. Census and the American Chamber of Commerce Researchers Association. With 73,079 observations from these data sources, the researchers used regression analysis to test the hypothesis of the potential associations. The results positively revealed that increased availability of chain supermarkets was statistically significantly associated with lower adolescent BMI and overweight among US adolescence. Similarly, Walker (2010) and his colleagues applied the approach of counting the number of supermarkets or groceries stores within a Census tract and calculating the number per resident in order to describe the food environment and came up with a similar positive association. This kind of density measure suffers a limitation in that it does not measure the actual access and lacks of discussion of transportation aspects.

Some studies have applied GIS to measure spatial accessibility between households and supermarkets. Zenk (2005) and his colleagues applied analysis tools in ArcView software to calculate the distances based on the geographic coordinates of neighborhood centroids and the nearest supermarkets. Michimi and Wembley (2010) examined the odds of obesity statistically associated with the distance to supermarkets. Their data on supermarkets was obtained from the 2006 Census Zip Code Business Pattern classified by the North American Industrial Classification System (NAICS). Two NAICS codes (445110 and 452910) of supermarkets and other grocery stores (excluding convenience stores) were combined, and a separate supermarket category was created for their analysis. The two categories were selected because these two categories of establishments are the primary providers of healthy foods, while convenience stores do not (Michimi & Wembley, 2010). Michimi and Wembley (2010) used the centroids of the ZIP Code Tabulation Area (ZCTA) to represent the population data in the analysis. The results showed that the odds of obesity prevalence increased with increasing distance from supermarkets in metropolitan areas.

Another widely cited study was done by Giang (2008) and his colleagues. They applied the Spatial Analyst tool in ArcView 3.2 software to create a supermarket density map and distribute the overall sale volume to a 1-mile distance catchment area and then divided the density map by a raster layer of population. Then, they compared the density map with supermarket sales and population data to the odds of diet-related disease prevalence. The results showed that where the availability of supermarkets was more limited, the residents suffered higher health risks in which diets act as a factor.

Based on its definition, food deserts are a topic that relates to spatial accessibility, and the travel mode of households has a profound effect on spatial accessibility, which some studies took into consideration in the analysis of food environments (Bader et al., 2010). Specifically in the U.S., the availability of vehicles is the most dominant factor of whether a household can access nutritious food. Based on a survey by the U.S. Department of Agriculture in 2009, 2.3 million, or 2.2 percent of the overall households in the U.S., live more than a mile from a supermarket and do not own a vehicle. Recent research has found that the pattern of accessibility is largely influenced by the availability of access to a private vehicle (U.S. Department of Agriculture, 2009). A number of studies have found evidence that for the disadvantaged households who live within food deserts without vehicles, the situation could be worse when having to expend more time or money to eat a healthy diet, so they may turn to unhealthy diets such as fast food (Bader et al., 2010; Larsen & Gilliland, 2008), which can lead to a higher BMI and a series of diet-related co-morbidities, such as obesity, diabetes, and cardiovascular disease (Beaulac et al., 2009; Bader et al., 2010).

2.2.2 Social Gathering Places

Social gathering places can be broadly defined as "a subset of the infrastructure sector and typically includes assets that accommodate social services" (New Zealand Social Infrastructure Fund, 2009). Mitchell (1995) claimed that public space represents the material location where "the social interactions and political activities of all members of the public occur." In the present research, the scope of social gathering places is limited to the facility locations that accommodate community gathering, such as churches, libraries, and local community centers.

The notion of creating public gathering spaces in the urban design practice is very old. Though not using the same term, there is also much planning and design literature that emphasizes the importance of access to gathering space from different perspectives. For example, Talen (2000) proposed that access to public services promotes "resident interaction" and "place attachment." She also reviewed literature related to social and environmental movements and concluded that access is a critical component of urban form due to the ability to promote the integration of activities based on research from Jacobs,

Lynch, Duany and Plater-Zyberk.

A number of researchers have demonstrated the importance of potential social gathering spaces in community place making. Egan (2004) proposed that the availability of these services and facilities promotes social interaction within the communities and help hold the communities together. Casey (2005) also proposed that these facilities of this category "enhance the quality of life, equity, law and order, stability and social well-being through community support." Although there is a lack of evidence that directly associates built environments with the level of social capital based on the present literature review on social capital, it can still be argued that social gathering places still play an essential role because they provide "the framework in which a community transacts economic, social and environmental activity" (Casey, 2005, p. 7).

The British Property Foundation (2010) pointed out that social gathering spaces are more than essential services such as hospitals. Public gathering spaces such as churches and community centers are important for the sustainability of communities in the long term as well. These public spaces are all crucial components of communities but are often overlooked. Van Bergijk et al. (2008) argued that neighborhood facilities such as libraries and cafes increase the chance for residents to meet each other, and such encounters are a vital precondition of social cohesion. Lotfi and Koohsari (2009) also concluded that access to these public spaces provides a venue for chance encounters and the framework of social capital. A lack of access to public interaction and gathering may hamper the social interactions and activities of all members of the public and lead to social isolation.

2.2.3 Sports and Recreation Facilities

Recreational facilities have been widely believed to play an important role in the process of children's cognitive development. In addition, recreational facilities enable children and teenagers to engage with the surroundings (Smoyer-Tomi et al., 2004). Similar to social gathering places, sports and recreation facilities are also considered as important neighborhood resources that contribute to social interaction and create attachment to a community (Connor & Brink, 1999; Casey, 2005; Van Bergijk et al., 2008).

2.2.4 Healthcare Facilities

The fourth category is healthcare facilities, which are widely considered a vital sector of services (Rosero-Bixby, 2004). Accessibility to medical clinics has been an issue in need of attention in urban planning (Ngui & Apparicio, 2011; Lasser et al., 2006). It is argued that poor access to medical clinics and other healthcare facilities may result in health problems due to the tendency of people with simple health problems choosing not to consult healthcare professionals and consequently degenerating into more complex diseases (Ngui & Apparicio, 2011). *The Canada Health Act* (CHA) adopted in 1984 emphasizes the significance of access to healthcare facilities, demonstrating that all Canadians should have access to necessary medical services without restrictions, and that the Canadian healthcare system has the obligation to make sure healthcare facilities are accessible to all citizens.

2.2.5 Parks

The idea that exposure to the natural environment has health benefits is not new and can be found in various cultures (Ulrich, 1979). American landscape architect Frederick Law Olmsted believed that natural environment has positive influences on city dwellers by bringing them "tranquility and rest to the mind" (Ulrich, 1979). In the past two decades, there has been growing interest in the benefits of exposure to nature regarding well-being, and it is widely believed that there is a positive relationship between the two (Herzog & Strevey, 2008; Groenewegen et al., 2006). The mechanism is usually explained through the effect of a natural environment possibly helping to reduce stress and mental fatigue (Groenewegen et al., 2006). One of the most influential laboratory experimental studies was conducted by Ulrich (1979), who collected evidence on health benefits by showing participants photographs slides of urban or natural environments and testing the participants' emotions and anxiety state. The results illustrated that the exposure to nature scenes led to lower anxiety levels compared with the urban scenes and supported the notion that providing natural environments in urban areas could be beneficial in psychological aspects beyond aesthetics value. Kaplan (1995) explained the beneficial effect from the perspective of attention restoration, which proposed that contact with a natural environment could alleviate fatigue and thus improve attentional function.

More recently, some epidemiological research has shown mounting evidence that there is a statistically important relationship between the natural environment and perceived health indicators in a large population sample (Groenewegen et al., 2006). For example, De Vries (2003) and his colleagues collected a large set of information on the general health of 300,000 people in 2001 to make sure that there were sufficient samples among the population and created a dataset with indicators of the health and well-being of the target group. Then, they applied GIS techniques to associate the health and well-being data with land use data (for example, the availability and amount of green space within a certain distance of an individual's home). The results of their data analysis supported a number of hypotheses on the effects of the natural environment and provided evidence of positive association between the amount of green space and people's health and well-being outcomes.

2.3 Nature of the Topic of Services Distribution

The arrangement of spatial distribution of public services and resources is paramount for planners and policy makers, and the design of the policy always raises both efficiency and equity issues (Dietz & Atkinson, 2010; Bleichrodt et al., 2004; le Grand, 1990). The former is usually about the benefits and the costs. At the same time, the policy makers also consider how the certain resources are distributed among people, which is related to equity (Litman, 2003; Talen, 2012). Both of the objectives of equity and efficiency have been regarded as goals of vital importance in urban planning and welfare policy making (Talen, 2012; Le Grand, 1990).

In the field of urban planning and services distribution, generations of researchers have explored a wide range of theories and made contributions to the discussion of the concepts of equity and efficiency (Blanchard, 1986; Le Grand, 1990; Dietz & Atkinson, 2010). However, despite the continuing research interest in this field, the terms "equity" and "efficiency" remain ambiguous and inconsistent in previous research. Researchers may tackle the issues from various angles and mean different things when they refer to these concepts (Blanchard, 1986; Hay, 1995; Litman, 2002; Le Grand, 1990). Due to the unresolved disagreement on the basic concepts of equity and efficiency, the question of how social services can be considered as equally or efficiently distributed remains far from solved (Talen, 1996; Kaufman, 2007).

The performance measures of the equity and efficiency of a social-services and welfare system are developed in very different ways based on the various focuses of the research. In addition, the definition and analysis of equity in transportation planning is particularly difficult, because there are many dimensions to consider (Litman, 2003). As a result, it is of vital importance to first clarify and define these fundamental concepts. It needs to be noted that by pointing out the inconsistency of the definitions of equity, the intent is not to propose an alternative notion for these terms that can be applied universally in multiple research topics. Instead, the aim is to examine a set of definitions of the terms based on various beliefs within a framework that is explicit for further discussion.

2.4 Notion of Equity

The achievement of equity in the distribution of public services and resources has been considered as a goal of vital importance among planning studies (Talen, 2012; Litman, 2002; Litman, 2003). The discussion is fundamentally related to "who gets what and by what rules" (Blanchard, 1986). As a result, equity analysis is unavoidable in the public service policy-making process, and the equity concerns act as a key issue in the assessment of spatial distribution of desired services (Jacobson et al., 2004; Litman, 2003; Litman, 2002; Talen, 1998; McMaster et al., 1997; Blanchard, 1986). In the field of transportation and urban planning, the equitable distribution of public resources is a complex topic. It involves not only the methodology of measuring people's ability to reach the services, but more fundamentally, it is rooted in the theoretical foundation of social justice and value judgments. As a result, before answering the question of how equitable the distribution should be and proposing an analysis model, it is necessary to review the brief background of the debate on equity and clarify all the competing definitions of the concept of equity from various theories as well as its measures, especially for the topic of services distribution in the context of urban and transportation planning.

Generations of researchers have made contributions to the discussion of the concept of equity or its equivalents fairness or justice, and have explored a wide range of theories (Litman, 2003; Blanchard, 1986). There are many transportation and urban planning studies that have discussed the concept of equity and contributed to the body of knowledge on this topic (Hay, 1995; Litman, 2002; Karner & Niemeier, 2013). For example, Blanchard (1986) proposed that equity is "a perennial issue because virtually every public policy has benefits and costs that are distributed more or less unequally among citizens." More recently, Bourguignon (2007) and his colleagues proposed that equity be defined in terms of two basic principles: (a) equal opportunities and (b) avoidance of extreme deprivation in outcomes. However, despite the rich literature, the questions of how urban services should be distributed equally among citizens and how the level of equity can be measured have not been fully explored in detail (Kaufman et al., 2007).

The intended meanings of "equity" are not necessarily consistent, and there is no consensus about what an equitable distribution of services could be among previous researchers. The definition and analysis of access equity in urban and transportation planning is particularly difficult because there are numerous aspects to consider and various ways to interpret the concept of equity (Litman, 2002). So, the first step is to discuss the complexity of the definitions of equity and how they can be taken into account in the present research topic, which is urban services distribution and transportation planning policy. Considering this framework of definitions helps to provide a comprehensive understanding of how these definitions are related to certain research and how they can be applied in certain circumstances, it is useful to define and justify the specific notion of equity adopted for the present research in context of urban services policy and transportation planning.

First of all, considering that the notion of equity is always related to comparison among people, it is important to analyze the relevant characteristics that distinguish groups by need or ability, such as social class or transport ability (Litman, 2002). Furthermore, in order to measure the level of equity in the distribution of certain goods or services among social subgroups, it is crucial to associate the spatial distribution of the goods with the spatial distribution of population subgroups that vary in need and ability (Talen, 2011).

In order to explore the characteristics of a population receiving urban services, the concept of social stratification must first be considered in the discussion of the notion of equity and the interrelationship between the social stratification, along with social equity/ inequity. The concept of social stratification, which refers to "unequally distributed resources within a social system" according to Erikson and Goldthorpe (1992), includes layers that mainly relate to the individual characters of two dimensions: (1) social-demography and socio-economy factors such as gender, income, and social class; and (2) lifestyles and personal attitudes (Bergman, 1998). The present research focuses on the former.

To discuss the social demography and socio-economy dimensions of social stratification, the notion of "inequity traps," is first introduced (Bourguignon, 2007). Conceptually, the notion of inequity traps is tightly related to the first dimension of social stratification. Bourguignon (2007, p. 243) described a social phenomenon in which the difference in a social service distribution among socio-economic groups could continue to exist while the disadvantaged groups do "persistently worse" than some other groups even all social groups are equally considered. The service here includes multiple distributions of opportunities and services. One example mentioned is the public school system. Assuming that the quality of teaching is related to the teaching staff, whose wages are influenced by the capital market, a richer family would pay the fees to send their children to private schools which provide higher quality of schooling, while children from poorer families would have to go to free public schools since they cannot afford the fees. If the policy makers do not value the public school from a budget perspective, the quality of public schools might be hampered. In this case, the children of the poor might be more likely to stay poor because they receive poor schooling, while the children of the rich could stay rich because they would be able to attend better schools. In other words, the quality of the schools reflects the institutional differences in economy and politics of the parents. Based on the discussion, Bourguignon (2007, p. 236) defined "inequity traps" as "persistent differences in power, wealth and status between socio-economic groups that are sustained over time by economic, political and socio-cultural mechanisms and institutions."

Based on different attitudes towards population groups, Litman (2012) distinguished the categories of equity as Horizontal Equity and Vertical Equity. Another review came from Martens and Golub (2012). Litman's theory on equity has been of vital significance in the discussion of the notion of equity. Litman (2002) also addressed that there are two principles of equity in transportation planning in general, which are Horizontal Equity and Vertical Equity. Though he did not put it clearly, he distinguished these two based on attitudes toward the potential differences among different social groups.

2.4.1 Horizontal Equity

The concept of Horizontal Equity was developed based on the concept that equal individuals and groups should receive equal shares of resources (Litman, 2002; Talen, 2012). Blanchard (1986, p. 33) referred to this kind of equity as "Strict Equality," which is defined as an "equal share for everyone." This category of equity was also referred to as "equitable distribution" (Crompton & Wicks, 1988, p. 288; Talen, 1998, p. 24). More recently, Bourguignon (2007) described that equity should be defined mainly based on equal opportunities. In the field of urban-services policy and transportation planning, Talen (1998, p. 24) used the term "equality," in which "everyone receives the same public benefit, regardless of socioeconomic status, willingness or ability to pay, or other criteria; residents receive either equal input or equal benefits, regardless of need." In the field of planning, planners applied distributive policy based on this category of equity criteria partly because of its simplicity. For example, resources can be distributed according to per capita allocation without consideration of differences of neighborhood groups (Talen, 1996).

2.4.2 Vertical Equity

As mentioned, the other principle of equity is referred to as Vertical Equity (Litman, 2002). As one of the pioneers of equity planning research, Krumholz (1982, p. 166) pointed out the inherit unfairness nature in the urban development process and proposed that the nature of equity of planning is an effort to provide more "choices to those... residents who have few, if any choices." Following this thread, Krumholz (1982, p. 163) proposed the notion of "equity planning," which was intended to mitigate the inequity related to social class distinctions. Crompton and Wicks (1988, p. 288) proposed the term "compensatory equity," meaning that equity in the distribution of public resources needs to respond to the need of groups. The specific interpretation of "need" may vary according to the services and targeted groups. Talen (2002) contended that equity is defined in relation to the "spatial locations of population subgroups," which implied that equity is based on need. The analysis of equity distribution based on need can be interpreted as linking spatial service facilities with target populations who need access to them (Talen, 1996). Lucy (1981, p.448) referred to this kind of equity as "unequal treatment of unequals." She further argued that this "unequal treatment" should be based on factors tightly related to the nature of the benefit (Lucy, 1981, p.449).

Cohen (1989, p.908) further built on his argument and distinguished strong and weak equity arguments regarding what "ought to be equalized." The strong argument indicates that the welfare or resources should be distributed as equally as possible for the people, while the weak argument claims that the welfare or resources should be distributed as equally as reasonably possible in some dimension but have certain limitations related to other values that also need consideration. In the practice of transportation and urban planning, equity is subject to possible limitations due to its complexity and the impossibility of complete equality, which implies a choice for a weak equality (Martens, 2012; Cohen, 1989).

Based on this discussion, in the context of urban service distribution planning, the notions of vertical equity and weak equality are adopted, and **Social Equity** is defined in the present research as individuals and groups that differ in need and ability being treated differently in order to be provided with equal opportunities to resources. At the same time, the notion of **Social Inequity** can be defined as access to resources among subgroups of citizens not being equally distributed.

2.5 Notion of Accessibility

In order to answer the questions of how equal access to spatial distribution of urban services can be achieved, we first need to clarify the concept of accessibility and answer how accessibility can be measured. Accessibility is a concept that has widely been considered as an important characteristic of geographic space and in transportation planning, urban planning and landscape architecture in the last 50 years (Geus et al., 2012; Achuthan et al., 2010; Comber et al., 2008). Improving access to crucial urban services in urban environments has become a significant goal in transportation planning and policy making processes (Geurs et al., 2012). Hansen (1959, p. 73) first proposed the definition of accessibility as "the potential of opportunities for interaction." Weibull (1980, p. 54) considered accessibility as "a property of opportunities for spatial interaction." More recently, Handy and Clifton (2001, p. 68) argued that accessibility has been receiving increasing research interest from urban planners because it "reflects the possibilities for activities, such as working or shopping, available to residents of a neighborhood, a city, or a metropolitan area." Luo and Wang (2003, p. 867) defined accessibility from a geographic stand of point as "the relative ease by which the locations of activities, such as work, shopping, and health care, can be reached from a given location" in the field of geography. Litman (2003, p. 2) proposed that accessibility essentially describes an individual's ability to reach desired goods, services, activities and destinations — collectively, "opportunities."

In a more recent paper, Litman (2008, p. 5) highlighted the concept that accessibility is a "general concept used to describe the degree to which a product, device, service or environment is accessible by as many people as possible." He also discussed access as "connections to adjacent properties." Levinson and Krizek (2008, p. 7) stated that accessibility measures "the ease of reaching activity destinations." Though not using the same term, Leslie (2007, p. 113) and his colleagues described accessibility as "directness of the pathway between households and places of destinations and is based on the design of the street network." It has also been pointed out that in the context of geography, accessibility in an efficient connectivity network is different from accessibility in regard to relating points in a network (Levinson & Krizek, 2008). To conclude all these definitions and methodological thoughts, the notion of accessibility is defined here as **"the ability of people reaching possible benefits or locations within a system"** within the context of geography and transportation planning in general.

Reggiani (2012) introduced an analytical form for the notion of accessibility as follows:

$$Acc_i = L_j \cdot D_j \cdot f(\beta, c_{ij})$$

where Acc_i defines the accessibility of a node or zone i, D_j is a measure (or weight) of opportunities and activities in j, and $f(\beta, c_{ij})$ is the impedance function from i to j, with β as its cost-sensitivity parameter.

Based on these theoretical definitions and the debate on accessibility, three elements in this definition can be found: a social-demographic dimension, land-use/spatial opportunities allocation, and travel option/transportation infrastructure that connects the previous two elements (Geurs & van Wee, 2004). The result is the notion of accessibility as a three-fold framework which consists of these three dimensions. The travel option component represents the travel time/distance between the locations of the activities, which is affected by the quality of the network and certain characteristics of the target groups. The spatial location of the destination component means "the amount and the spatial distribution of the demand for or supply of opportunities (Geurs & van Wee, 2004, p. 129)." The social demographic component indicates the social stratification of the population, including income, age, and other relative characteristics of the individuals that affect the need and level of accessibility.

2.6 Measures of Accessibility

Urban planners and policy makers have studied the evaluation of the spatial distribution of urban services for several decades (Algert et al., 2006; Apparicio et al., 2008; Arentze et al., 1994; Bader et al., 2010; Beaulac et al., 2009; Brownson et al., 2009; Chin et al., 2008; Comber et al., 2008; Geurs et al., 2012; Handy & Niemeier, 1997; Handy & Clifton, 2001; Higgs, 2005; Iacono et al., 2010; Kelly et al., 2010; Martens, 2012; Neutens et al., 2010; Ploeg et al., 2009). As the spatial pattern of services distribution among groups in an urban area has always been constantly changing, this nature of accessibility analysis requires urban planners and policy makers to develop measurement tools to identify poor levels of access and evaluate how proposed improvement meets planning objectives (Arentze, 1994). The measurement methodology that they used has been a critical part in the process of evaluating whether the access to the urban services is equitable among social groups (Neutens et al., 2010).

There are many different ways to establish the framework of measures of accessibility that vary in terms of perspective and parameters (Achuthan, 2010; Talen, 2011). What these measures examined includes a wide range of dimensions. At one end of the spectrum, accessibility measures focus on people's satisfaction and received welfare based on utility theory (Mladenka, 1980). At the other end of the spectrum, accessibility is interpreted as the spatial allocation of services (Knox, 1978; Talen, 2001). The application of these measures eventually determines how accessibility is measured.

Neutens (2010) and his colleagues conducted a thorough review on accessibility measures used by urban planners and transportation researchers in the context of public service distribution. Their discussion provided a methodological comparison among accessibility measures. They distinguished various accessibility measures into two major groups: **Place-based Measures** and **People-based Measures**.

Based on focuses on a particular dimension in equity analysis, Martens and Golub (2012) developed welfare-resources-midfare theory. Their discussion was in line with Dworkin's (1981) theory, which divided the concept of equity into two categories: **Equality of Welfare** and **Equality of Resources.** This dichotomy, together with a third category later developed and described as **Equality of Midfare** by Cohen (1989), is helpful for clarifying the long-standing debate on the topic of spatial equity analysis, and is thus extremely valuable for the choices and development of different potential equity measures (Martens & Golub, 2012).

The present discussion combines both of the reviews and examines the characteristics of the accessibility measures they have covered based on the two mentioned categories: analysis objects (place-based or people-based) and the focus of analysis (welfare, midfare, or resources). The aim is to provide a framework for further discussion on the relationship between the measures and equity theories in the context of urban and transportation planning. The conclusion would facilitate the choice and the implementation of accessibility measures.

2.6.1 Place-based Measures and People-based Measures

Based on the review by Neutens et al. (2010), the first group of accessibility measures is referred to as place-based measures, which is a conventional approach to measuring accessibility that has been applied by many early researchers (Neutens et al., 2010). Generally speaking, this kind of measure examines the proximity from key locations (such as homes) to desired service locations (Miller, 2007). Many previous researchers have applied various place-based measures. Neutens et al. (2010) provided a systematical review of place-based measures based on the framework first proposed by Burns (1979) and extended later by Miller (1999).

a. Place-based Measures

The simplest way of performing place-based measure would be a route calculation between an origin and an urban service destination (Achuthan, et al., 2010). This particular kind of measure is referred to as distance measures (Neutens et al., 2010). Distance measures focus on the calculation of travel distance or time between an original point and a destination as a means of evaluating the level of accessibility (Martens & Golub, 2012). More specifically, two kinds of measurement approaches fall into this description: DMIN and TMIN (Miller, 1999). DMIN and TMIN respectively measure the network distance and the travel from an original location (usually an individual's home location) to the closest service location within transportation infrastructure. In these two measures, the level of accessibility is considered lower if the service location is further away from where an individual lives (Neutens et al., 2010). The difference between the two is that TMIN takes the travel mode of the individual into account.

Another simple type of place-based measure is infrastructure-based measures, which assess the level of services within the context of a transport system. They are usually applied to a transport network and compare the level of resource availability for individuals. The most common approach is to sum the total spatial opportunities available within a certain geographic area, which are usually Census tracts or neighborhoods (Arentze, 1994).

The third type of place-based measure is cumulative opportunities measures (CUM). Cumulative opportunities measures do not only consider the closest services, but all alternative opportunities within a specific cutoff travel time or distance (Neutens et al., 2010). Cumulative opportunities measures calculate the total number of accessible opportunities within a given cutoff travel distance or time. This type of measure can be expressed by the following formula:

$$CUM = \sum P(t_{hp}),$$

where

$P(t_{hp}) = 1$, if $t_{hp} <$ cutoff

$P(t_{hp}) = 0$, otherwise

The fourth type of place-based measure is gravity-based measures (GRAV), which apply similar approaches of calculating potential opportunities as the cumulative opportunities measures (CUM). But gravity-based measures use a distance impedance instead of a fixed travel distance. Besides the spatial allocation of resources, both of these measures take some characteristics of people into account related to their mobility within the transportation network (Martens & Golub, 2012).

Gravity-based measures can be expressed by the following formula (Neutens et al., 2010):

$$GRAV = \sum Ra_q \exp[-\lambda_m \min(t_{hp})],$$

where

a_q is the overall attractiveness of service locations q

λ_m is the distance parameter of mode m.

Besides the spatial allocation of resources, both the GRAV and CUM measures take some characteristics of the mobility of people within the transportation network into account.

b. People-based Measures

The second group of accessibility measures is people-based measures (Neutens et al., 2010). This group of measures not only the focus on transportation infrastructure and spatial services allocation, but also an individual's detailed travel behavior (Neutens et al., 2010). Based on the review by Neutens et al. (2010), the first type of people-based measure is space-time measures (STP), which is rooted in the work of Lenntorp (1978), who examined whether activities are available for individuals in an urban environment. The space-time measures do not focus on travel activities themselves, but on the people's capability to participate in travel and activities with the time-space constraints in a specific transportation system (Martens & Golub, 2012). These measures take into account people's home and activity locations and what is available within an individual's time budget, and then calculate the spatial distribution into a capability with the consideration of people's travel mode and limitations (Neutens et al., 2010).

The second type of people-based measures is the utility-based space-time measure. This type of measure applies a utility-based approach to space-time measures based on the work of Burns (1979) and Miller (1999). Different from the space-time measures derived from Lenntorp's (1978) theory, this type of measure reflects the desirability of opportunities that people derive from travel and activity participation, not merely the number of opportunities (Neutens et al., 2010). In other words, the utility-based space-time measures assess the maximized utility that one person can derive from participating in activities with certain space-time constraints (Martens & Golub, 2012).

The third type of people-based measures is utility-based measures, which are rooted in random utility theory. Random utility theory defines accessibility as a set of potential transport choices that maximize a person's utility (Geurs & van Wee, 2004). As a result, utility-based approaches measure the outcome as the level of accessibility. Containing characteristics of the transportation system as well as characteristics of people, utility-based measures represent a choice set based on an individual's decision process.

Though not summarized by Neutens et al., there is another type of people-based accessibility measure that is worth mention, which is the doubly constrained accessibility measures (Martens & Golub, 2012). The doubly constrained accessibility measures were developed based on the spatial interaction conceptual model, focusing on people's ability to transfer resources to welfare (Geurs & van Eck, 2001). In these models, the measures provide a sophisticated approach to balance the potential factors by integrating job-seekers' competition for jobs as well as employers' competition for job-seekers. However, these measures have deficiencies: they usually do not contain the link between skills of people to the particularities of jobs, which could be a fairly important factor in the analysis. In addition, they do not take the mobility of people into account, which is important in evaluating the ability of people to participate in activities.

2.6.2 Welfare-Resources-Midfare Theory

Martens and Golub (2012) pointed out that the development and choice of accessibility measures are closely associated with various focus on specific groups, which could be distinguished by geographic locations, ethnic groups, travel mode, or income. Different measures address the need for different policy making considerations. Therefore, the choice of accessibility measures that respond to certain resource distribution questions should be utilized based on a clear understanding of the equity distribution theory in the context of urban planning (Martens & Golub, 2012). Martens and Golub (2012, p. 195) argued that in order to develop a framework of equity-theoretic approaches for accessibility analysis, the predominant step is to explore and define the "focus or dimension" of an equity analysis in the context of transport and urban planning (the question here is what should be distributed in an equal manner). Martens and Golub (2012, p. 195) termed this specific "focus" as "equalizandum," which refers to the "central dimension of comparison" in the equity analysis.

a. Equality of Welfare

As mentioned previously, Martens and Golub (2012, p. 196) highlighted three categories of equity theories and approaches based on different equalizandums: equality of welfare, equality of resources and equality of "midfare." Equality of welfare, aims to achieve a way for welfare to be equalized through redistribution of goods and services. Litman (2012, p. 5) described this concept of equity as "equity of outcome." However, there are two distinguished groups of theories that differ in their ways of interpreting and conceptualizing the concept of welfare according to Dworkin's (1981) review.

The first group of theories was referred to as "success theories of welfare" by Dworkin (Martens & Golub, 2012, p. 197). These theories considered welfare as people's success in "fulfilling their preferences, goals and ambitions" according to Dworkin's argument (Martens & Golub, 2012, p. 197). A person's welfare is said to increase if their preferences are more fulfilled. The second group of theories interpreted welfare from the aspects of people's conscious life, such as pleasure or pain. In this case, people's welfare increases as they have more enjoyment. Martens and Golub (2012) claimed that these two groups of theories are comparable and similar to a great extent.

Overall, the equality of welfare involves a potential distribution of goods in which welfare is equalized. In other words, the level of people's welfare determines the way resources should be distributed. Particularly for transportation planning, the equality of welfare focuses on the "satisfaction" that people gain in travel activity (Martens & Golub, 2012, p. 198). This satisfaction could be derived from either the travel itself, or the benefits from the trip (for example, the activities they participate in on the trip, or the resources they gain from the trip).

The most important objection against equality of welfare is referred to as the "expensive tastes" argument proposed by Cohen (1989, p. 907). His main argument can be concluded as "one has expensive tastes and the other does not" (Cohen, 1989, p. 907). Equality of welfare calls for the same level of equal satisfaction by providing each person sufficient resources. This indicates that a person with more requirements, or "with expensive tastes" in Cohen's words, will need more resources than a person with modest tastes to be equally satisfied (Cohen, 1989, p. 913). As illustrated previously, the level of pleasure could derive from either the travel itself or the activity participation. So, if we transpose this expensive tastes argument to the field of transportation planning, one situation could occur in which two people may derive different levels of pleasure from activity participation. Applying the equality of welfare theory would imply providing better transportation services to the person who has a lower level of pleasure in travel in order to compensate for the pleasure activity participation to guarantee the equality of welfare. This conclusion from the application of equality of welfare seems unfair and goes against a normal understanding of equity (Martens & Golub, 2012).

Sen (1980) also contributed to the argument on equality of welfare by contending that the notion of equality of welfare is not suitable for policy making unless it is in a situation in which individuals can adjust their expectations. Translated to the field of transportation and urban planning, it can be interpreted as a person without a private vehicle having to accept a less convenient transit system and not calling for a better system. This conclusion obviously counters the common welfare approach and beliefs (Martens & Golub, 2012).

b. Equality of Resources

The second categories of equity theories is referred to as equality of resources by Dworkin, who helped the development of accessibility measures in transportation planning related equity analysis (Martens & Golub, 2012). Instead of welfare, Dworkin (1981) proposed resources as a more proper equalizandum. The notion of "resources" refers to "the sum of material resources and mental and physical capabilities" in his theory. He stated that people should have equal initial resources and thus need to be compensated for deficiencies. Litman (2002, p. 5) described this kind of equity as "equity of opportunities" that everyone deserves, indicating that disadvantaged people should have adequate access to social services and opportunities.

Applying this argument to the field of transportation planning, it indicates that the distribution of transportation-related resources, rather than welfare of people, deserves attention and should be the object analyzed (Martens & Golub, 2012). In other words, researchers should focus on the resources that facilitate people's travel to approach the activities and services, rather than the welfare that people derive from the travel (as equality of welfare theories claim). All these approaches based on equality of resources also implied a focus on potential mobility of people, which can be interpreted as the "ease with which a person can move through space" (Martens & Golub, 2012, p. 201). In addition, in Dworkin's theory, the deficiencies that need compensation are solely the particularities of people related to transportation, such as disabilities or ownership of vehicles.

One of the major critiques of this equity of resources theory and approach was proposed by Rawls (1981) and Sen (2011). Rawls argued that an equity of resources approach "does not take into account the differences between the recipients of those resources" (Martens & Golub, 2012, p. 202). And Sen (2011) pointed out that different people require different sets of resources to fulfill identical needs. As a result, the focus on transportation infrastructure and spatial accessibility in equality of resources theories may fail to connect to people's real needs and satisfaction gained from the travel. Furthermore, Sen strongly criticized the concentration on goods rather than "what goods do to human beings" and claimed that it could be very misleading to judge what people really have merely from their ownership of the services and goods (Cohen, 1993, p. 16). To apply his critique in the field of transportation planning, it is argued that the analysis approach based on equity of resource theories focus on people's ability to access the services without really knowing what the ability exactly brings to people. Actually, this critique proposes that the particularities of the person are the missing key component for the distribution analysis in the field of transportation.

Similarly, Kaufmann (2007) and his colleagues pointed out that one shared deficit of location-based accessibility in transportation planning is that the correlation between spatial displacement of goods and services with social stratification has not been fully explored and clarified. In other words, these theories and analysis approaches were built merely on the spatial distribution of services and opportunities, while the actual demands of the services were not analyzed (Talen, 2002). This category of theory measurements of equity usually assumes that all households and individuals share the same level of access to these services, or simply pay no attention to the potential differences (Weber & Kwan, 2003). This assumption may not always be valid, because differences relating to social-demographic and socio-economic characteristics, such as car ownership rate and age, largely affected form how the households and individuals access services (Ohnmacht et al., 2012). In addition, different subgroups reacted to the change of spatial distribution of services in different ways (Ohnmacht et al., 2012). As a result, it is doubtful to ignore the differences of social characteristics among households and individuals in the discussion of transportation equity.

Geurs (2012) and his colleagues also addressed a similar concern on the distribution-based accessibility measures. They claimed that these kinds of measures fail to reflect the diversity in individuals as well as in social groups and thus needed further improvement in developing the indicators. But they also pointed out that because of fast improvement in techniques, considerable progress has been made since the 2000s in developing more complex and detailed accessibility measures by facilitating sophisticated algorithms on the GIS platform, which has the abilities to store, manipulate, and analyze data from multiple geospatial databases.

c. Equality of Midfare

In order to solve the debate between equality of resources and equality of welfare, Cohen (1993) proposed a third potential equalizandum than lies between the resources and welfare based on Sen's capabilities theory. He used the term "midfare" for this new concept, which "is constituted of states of persons produced by goods, states in virtue of which utility levels take the values they do. It is 'posterior' to 'having goods' and 'prior' to 'having utility'" (Martens & Golub, 2012, p. 203).

The core argument of equality of midfare theory is to focus on people's condition in the abstraction of utility. It is argued that neither possessions of goods nor satisfaction from goods matter (Cohen, 1993). Instead, it looks for a comprehensive characterization that stands for the range of people's well-being states. For example, on the topic of food desert analysis, the equality of midfare measures would look into people's nutrition level instead of either food supply (as the equality of resources theory would suggest) or satisfaction from eating food (as the equality of welfare theory would suggest). It is argued that the satisfaction is only the evidence of people's well-being, while the goods themselves are only the causes of well-being (Martens & Golub, 2012).

If we apply the theory of midfare to the field of transportation and urban planning, we can define the midfare as the extent to which a person can take advantage of his mobility in participating in activities (Martens & Golub, 2012). We can find a broad spectrum of factors that can affect a person's ability to interpret the access to services into welfare. The two most significant factors are land use patterns and personal circumstances. Land use patterns can be easily connected to people's ability to interpret the goods into welfare, because the spatial distribution of various categories of services is one of the most dominant factors. What Martens and Golub (2012, p. 204) referred to as "personal circumstances" includes financial situation, education and skill level.

2.6.3 Interpretation for Accessibility Measures

As discussed previously, in the debate of equality analysis, Cohen (1993) distinguished three categories of equity theories based on three equalizandums, which are welfare, resources and midfare. Each category focuses on a different aspect of people's well-being and thus indicates a different type of equity analysis. Based on Cohen's (1993) theory that distinguished the resource-midfare-welfare equality theories, Martens and Golub (2012) proposed the following table to correlate the equality theories with a variety of accessibility measures in position.

Two types of accessibility measures can be easily related to the **Equality of Resources** approach: infrastructure-based measures and distance measures (DMIN and TMIN in Neutens' and his colleagues' review). Both of these types of measures tend to be applied to transport networks and used to evaluate the level of resource provision. As discussed earlier, it can be easily identified that both of these measures ignore the differences among people and lack of socio-economic and socio-geographic components in these models (Sen, 1982; Kauffman, 2007; Talen, 2011). From the standpoint of an equality of welfare argument, it lacks components that reflect the potential differences among people who receive these services (Sen, 1982).

Various accessibility measures can be linked to the **Equality of Midfare** approach. Among these measures, the cumulative opportunities measures (CUM) and gravity-based measures apply identical concepts. As mentioned previously, the cumulative opportunities measures calculate the total number of accessible opportunities within a given travel distance or time as the level of accessibility. Gravity-based measures apply similar approaches of calculating potential opportunities as the cumulative opportunities measures. But Gravity-based measures use a distance impedance instead of a fixed travel distance. Besides the spatial allocation of resources, both of these measures take into account the mobility characteristics of people within the transportation network (Martens & Golub, 2012). It can be concluded that the cumulative opportunities measures and the gravity-based measures are closer on the spectrum to equality of resource approaches, rather than equality of welfare approaches.

Author	Equalizandum	Application
Rawls	Goods	Amount of food received
Dworkin	Resources	Amount of food received, corrected for disabilities
Cohen	Midfare	Nutrition level acquired from food matched to nutritional needs of person
Sen	Capability	Health from diet which allows for important activities
Bentham/Mill	Welfare/Utility	Preference satisfaction or enjoyment (utility) derived from eating

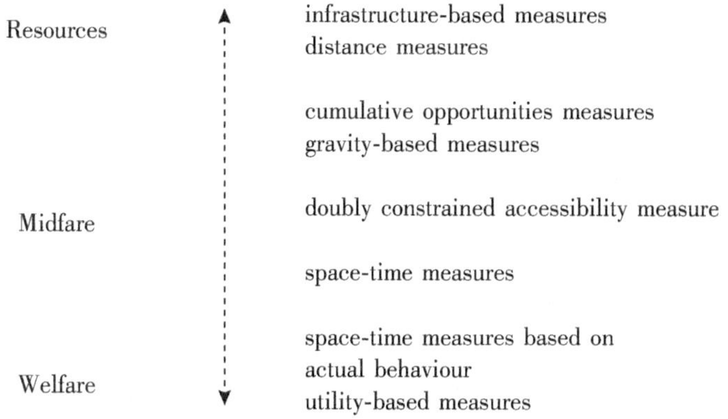

```
Resources    ▲    infrastructure-based measures
                  distance measures

                  cumulative opportunities measures
                  gravity-based measures

Midfare           doubly constrained accessibility measure

                  space-time measures

                  space-time measures based on
                  actual behaviour
Welfare      ▼    utility-based measures
```

Figure 1: Position of various accessibility measures in the resource-midfare-welfare sequence.
Source: Martens and Golub, 2012.

Another type of measures that fit into the equality of midfare approach is the doubly constrained accessibility measures (Geurs & van Eck, 2001). Doubly constrained accessibility measures job seekers' ability within the context of potential employees. It can be interpreted such that it focuses on people's ability to transfer resources (accessible job opportunities) into welfare (actual jobs). This concept fully captures the concept of equality of welfare theory.

The space-time measures can be also related to the equality of midfare theory because they do not focus on travel activities themselves, but on people's capability to participate in travel and activities with the time-space constraints in a specific transportation system

(Martens & Golub, 2012). Space-time measures take into account people's home and activity locations and calculate the spatial distribution into a capability, with the consideration of people's limitations. This approach actually evaluates the extent to which an individual can transfer resources (spatial allocation of services within certain transport network) into welfare with the individual's particular characteristics. This concept clearly resonates with the basic principles of the equality of midfare approach.

The final group of measures can be related to the **Equality of Welfare** theory, both of which focus on people's actual travel behaviors instead of potential opportunities. The first one is a utility-based approach. As the equality of welfare theory has proposed, utility-based measures are developed based on both characteristics of the transportation system as well as the characteristics of people. The second group of measures is utility-based space-time measures. As mentioned previously, these measures apply a utility-based approach to space-time measures by assessing the maximized utility that one person can derive from participating in activities with certain space-time constraints (Martens and Golub, 2012). Similar to the utility-based approach, these measures also follow the line of equality of welfare theory and focus on the utility that people derive from the travel and activity participation.

2.6.4 Conclusions

Based on the discussion from Martens and Golub (2012) on distinguished accessibility measures and their relationship with various groups of equity theories, we can draw several conclusions on the choice of accessibility measures development in the topic of spatial distribution of urban services (and more specifically, in a pedestrian network environment in the present study).

a. Based on the discussion of the notion of accessibility in the context of transport and urban planning, accessibility should be a three-fold framework, with the third component being the social dimension. Geus (2012) and his colleagues pointed out that one significant direction that we can identify in recent accessibility studies is addressing the population groups at various scales in the development of accessibility measures. Based on this argument, the conventional place-based approaches (or infrastructure-based measures), which lack a social-demographic component and do not take into account the differences among recipients groups, are not sufficient for equity analysis in the field of urban planning (Martens & Golub, 2012).

b. On the other hand, we can eliminate all accessibility measures based on the equality of welfare theory, since the utility that an individual derives cannot act as an efficient indicator for measuring accessibility in transportation planning. The reason has already been introduced in the section of the expensive tastes argument by Cohen's (1989) and Sen's (1980) argument against the equality of welfare theory in accessibility measures development. The equity analysis of urban service allocation should not be built upon the level of welfare people derive from the travel and activity participation (as the equality of welfare theory would argue), but should compare various groups by their possible travel behavior within a certain transport context.

c. Martens and Golub (2012) followed Cohen's theory and proposed using midfare as the better equalizandum. The midfare approach applies abstracted characteristics of people's well-being to stand for the indicator of people's condition and ability. However, the data that reflect people's ability to take advantage of the resources into welfare may not always be available or easily interpreted at the desired scale (Talen, 2012).

Instead, characterizing the **demand** based on socio-economic characteristics of population groups is proposed as the more proper equalizandum for equity analysis in transport and urban service distribution studies. Instead of finding a middle ground as a more proper indicator between the two ends (utility and resources) along the spectrum (as the midfare approach would propose), the **demand-based** approach links the two ends of the spectrum by matching the spatial distribution of services within a certain transport infrastructure context (the resources) to specific populations who need access to them to transfer into welfare (the utility). Unlike the equality welfare approaches, the actual travel behavior is not important in the demand-based measures (the critiques against the use of actual travel behavior have been illustrated). The potential access to categories of services matters. Talen (2011) also concluded that this set of need-based distribution measurement could be the most relevant in spatial equity research and has become a standard practice in geographic literature.

2.7 Defining Target Groups

In order to analyze the demand, the next procedure is naturally defining the target population groups. Despite the fact that the definition of target population has received extensive research interest in previous studies, no specific guideline has been provided (Karner & Niemeier, 2013). In most environmental justice studies, minority and low-income populations are identified as the target population, but not presumptively (Karner & Niemeier, 2013). Litman and Brenman (2012, p.14) distinguished the "demographic" characteristics (such as race) and "functional" characteristics (such as poverty and ability). They claimed that the "demographic" characteristics were often "ambiguous" in transport equity analysis and that measurement needs to be built upon "functional" characteristics (Litman & Brenman, 2012, p. 14). In addition, Karner (2012) stated that characteristics such as race, poverty, and ability may not always relate to the travel behavior of people. In a report to the US Congress, Ploeg (2009) and his colleagues proposed that individual characteristics related to travel mode, such as vehicle ownership, have a significant effect on the spatial accessibility measures of a food environment. Rose and Richards (2004) examined the relationship between a variety of measures related to food store access and household food use and concluded that the variable of vehicle ownership rate was statistically associated with healthy food consumption. Their study has provided supporting evidence that vehicle ownership rate is a statistically significant criterion for research on urban service accessibility. To conclude, considering the supporting evidence and the present analysis focus on pedestrian environments and behaviors, the vehicle ownership rate could act as a proper variable for target group selection in the Spokane metropolitan area.

2.8 Trade-Off between Equity and Efficiency

It is necessary to point out that equity is not the only objective in social welfare policy making. Another objective that appears high on the list of most welfare policies is related to urban service distribution (Le Grand, 1990). In welfare policy, the notion of Efficiency is usually defined as "the largest possible value of output given a production possibility set" (Bourguignon et al., 2007, p. 240). In his discussion on healthcare welfare policy, Bleichrodt (2004) and his colleagues claimed that when it comes to the allocation of urban services, people consider efficiency, which is the total amount, as well as equity, which is related to the distribution. In previous studies, when it comes to the discussion of equity and efficiency, it is widely asserted that there is usually a "trade-off" between the two in real-world situations. In other words, an implementation of policy that aims to "increase" one may result in a "decrease" for the other (Le Grand, 1990, p. 554). As a result, in the policy making process, a trade-off is highly likely between the two objectives (Le Grand, 1990; Dietz & Atkinson, 2010).

In these situations, the goal of policy making is not just to maximize the potential benefits for the maximum number of members, but also to pay additional attention to the least advantaged members and provide them access to the benefits. In some cases, it is considered to be more ideal for the whole community to exchange some efficiency losses for

2.8 Trade-Off between Equity and Efficiency

an economic equilibrium. This trade-off notion can be dated back to contract theory. Generally speaking, contract theory addresses how all different parties can construct contractual arrangements in policy making. It is proposed that all members in a community have a contractual liability to each other. Contract theory provides a theoretical background for the link between equity and efficiency. Contract theory is widely applied in the decision making process of certain utility structures. In order to achieve an optimal final decision, an optimization algorithm is usually created to determine the optimal decision (Bolton & Dewatripont, 2005).

Le Grand (1990) summarized two types of situations that can be referred to as "equity-efficiency trade-off" in social welfare research. The first one is called "production," which is related to the "possibility or feasibility of alternatives" (Le Grand, 1990, p. 556). The second type of trade-off is "value," which relates to the "extent" to which one of the objectives can be traded off for another objective. This kind of equity-efficiency trade-off is consistent with the consideration in resources allocation (Le Grand, 1990, p. 555). In the context of resources allocation, most situations that policy makers need to deal with are "highly efficient distributions but grossly inequitable," along with distributions which are "extremely equitable but highly inefficient" (Le Grand, 1990, p. 555). Le Grand (1990) also proposed reducing the two objectives into one, and that the two objectives can be substituted for one another. By combining the two objectives into one criterion, the best among all mixtures can be potentially implemented in the evaluating and decision-making process.

The trade-off can be illustrated with the following figure. X_I and X_{II} indicate the two objectives of equity and efficiency. The convex lines are the welfare contour lines for a policy maker to evaluate. The points on each contour represent all the possible combinations of equity and efficiency. The concave line represents the potential productivity regarding the two objectives given the constraints. The points on the concave line represent the maximized value of one index with a given amount of the other (Le Grand, 1990).

To apply the theory to the present research, the preferred model should be able to incorporate equity analysis into a cost-utility analysis to allow for a trade-off between efficiency and equity with the given amount of resources for prioritization of the improvement in infrastructure.

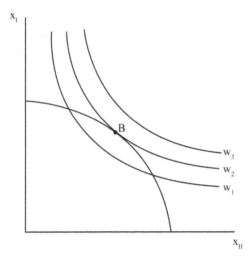

Figure 2: Trade-off between Equity and Efficiency.
Source: Le Grand (1990).

2.9 Geographic Information System (GIS) Platform

Recently, GIS has been widely used in transportation and urban planning and provides an ideal platform for accessibility research (Talen, 2011). By allowing researchers to collect, input, manipulate, and analyze complex spatial datasets related to population settlement and urban environment, GIS enables the operation of spatial measures of the built environment, and provides spatial linkage that enables the measurement of environmental factors and visual representation (Yigitcanlar et al., 2007).

Talen (2011) proposed that the power of visualization can act as an effective tool in the decision making process of resource distribution issues. She referred to this concept as "GeoVisualization," which is defined as "a set of tools and techniques supporting geospatial data analysis through the use of interactive visualization" (Talen, 2011, p. 458). The geovisualization of access to resources relies on the techniques of GIS and maps the pattern of accessibility in a straightforward way by revealing the level of access visually. More importantly, it contributes to the public discussion in an interactive approach, in which potential changes can be shown in real time with changes in input variables (Talen, 2011).

Another concept to incorporate in the model is "GeoDesign." The theory of geodesign proposes that the computer-aided design models of a current stage act as representational tools, which are used to visually represent either existing conditions or proposed designs (Dangermond, 2009). Similar to traditional drafting-board and manual drawings, their functions are limited in drafting and documenting. In urban planning and landscape architecture, typical examples include network analysis, view shed calculation, buffer zones, etc. Currently, we are at a digital threshold of the evolution to the next stage of computer-aided design tools into simulation tools. The paradigm shift from representation to the next stage means that computer-aided design technology would be involved in the thinking process of design more actively by providing quantitative assessment tools for different design interventions in real time and tremendously optimize the decision making process.

CHAPTER THREE   METHODOLOGY

In this chapter, the methodology of this research is first clarified from the larger domain of the conceptual framework of research methods. Then, in the rest of this chapter, the methodology of how to collect essential datasets and establish the GIS-based mode is described. In addition, the step-by-step methods of calculating residents' pedestrian access to a variety of categories of Life-Needs Service Facilities is illustrated within the Spokane PTBA, along with methods for determining the importance of each missing sidewalk in prioritization analysis. Section 3.2 describes data collection procedures, including the creation of pedestrian network model, collection and categorization of the Life-Needs Service Facilities dataset. Section 3.3 describes the overall data analysis procedures and explains the methods that were used to evaluate the accessibility and prioritize the potential improvements based on equity and efficiency criteria.

3.1   Theoretical Framework

In the history of architecture and planning, a great variety of research activities has been conducted. Because the field of architecture and planning covers a diverse array of topics, the scope of architectural and planning research has encompassed a broad range of paradigms. Therefore, it could be challenging to cover such a diversity of research concepts and methods (Groat & Wang, 2002).

Groat and Wang (2002) established a conceptual framework for understanding the multiple systems of inquiry. To begin with, they argued that the traditional quantitative and qualitative terminology is not suitable for architectural research because it overemphasizes the tactic level and entails certain ontological and epistemological assumptions. They also claimed that the continuum model by Morgan and Smircich focuses on particular strategies or tactics, rather than a system of inquiry. Instead of these two frequently used models, they proposed the notion of three paradigmatic clusters: postpositivism, naturalism, and the emancipatory paradigm. At the same time, there are many different research strategies and tactics that can work in consistent ways with the chosen paradigm. These strategies and tactics include: interpretive-historical research, qualitative research; correlation research, experimental and quasi-experimental research, simulation and modeling research, logical argumentation, and case studies.

Based on their conceptual framework, the present research methodology can be situated as logical argumentation research in the emancipatory paradigm. First, equity and efficiency were selected as the two criteria for prioritization, and the concept of equity was not considered a value-free objectivity, but as complex, socially constructed realities. The reality of dynamics among the inquirer, the planning policy makers, and the citizens should be recognized. The multiple realities are also the initial motivation of the customized prioritization tool proposed in the present model, with which the users can easily modify the weights between the two criteria to accommodate different social values. Second, the present research strategy falls in the category of logical argumentation research, because in this study, a framework (the GIS-based analysis model) is organized with quantitative measures that tie together different factors (pedestrian accessibility, equity and efficiency) and establishes the Life-Needs Service Facilities category as a basic feature in the argument. These features are resonant with the basic characteristics of logical argumentation research.

3.2 Data Preparation Procedure

The objective of this dissertation is to create a GIS-based analysis model with measures that can be used to assess the importance of the sidewalk segments for pedestrian accessibility and prioritize the potential improvements for the sidewalk network. Among the datasets in this study, there are three key components: the pedestrian network dataset, a point features dataset containing locations of Life-Needs Service Facilities, and the socio-demographic dataset.

3.2.1 Pedestrian Network Creation

The Spokane Transit Authority (STA) obtained funding from a Federal Transit Administration (FTA) grant through the Job Accessibility Reverse Commute (JARC) program. The establishment of the program aimed to improve the transportation services for low-income persons to access jobs and other employment-related services, as well as commute services in the Spokane Metropolitan Area.

This implementation of the JARC program in Spokane included a project that was named LIFTS (Lifeplan Improvement through Feasible Transportation Services), which

CHAPTER THREE METHODOLOGY

collected the city's accessibility-relevant data and established a GIS database to assist low-income individuals to identify available job locations along with bus routes, potential housing options, and other supporting services' facility locations by integration in one GIS tool.

The critical part of the LIFTS project is the Spokane Pedestrian Network (PNET) model, which is a representation of the small-scale pedestrian network system (PNET). PNET provides detailed and accurate information about all potential pedestrian pathways and facilities, including the availability of sidewalks, ramps, marked road crossings, and potential barriers in the Spokane PTBA. The pedestrian accessibility barriers were determined by the standards of Americans with Disabilities Act (ADA) of 1990.

The creation of the Spokane regional pedestrian network map required extensive field and lab work. The foundation of the pedestrian network (PENT) was the Pedestrian Path Dataset. The Pedestrian Path Dataset was created from the road centerline dataset obtained from the Spokane County GIS department. Half the width of a typical street in Spokane is 16.5 feet. From these centerlines, a distance of eighteen feet is measured in both directions in order to represent the approximate location of the lines of sidewalks of these streets. These created line features were extracted from the layer of the original road centerlines and then labeled as pedestrian rights-of-way. This process was done automatically with the platform of ArcGIS. Then, the network was realigned with a City-of-Spokane (only) off-street (OS) paving layer that was developed from a grayscale orthophoto of Spokane in 1992. Further examinations were based on a comparison with the aerial photographs from Avista in 2012 after the original network dataset was created. The presence or absence of pedestrian pathways as well as their visible attributes were recorded manually in the lab to match the aerial map. Different types of path types, including sidewalks, non-formal paths, driveway crossings, road crossings, etc., were identified and recorded as line features in the pedestrian right-of-way dataset, regardless of their accessibility. This pedestrian network was created to show the right-of-way condition for an average pedestrian who has a normal level of mobility. The default walking rate was set as 3 miles per hour.

The second part of PNET was a point dataset which recorded point barriers and pedestrian aids (such as curb ramps). This dataset contains the locations of features that could be barriers for mobility-impaired pedestrians, including planters, street signs, changes of surface height, and non-ADA-compliant ramps. There are several criteria to determine whether a feature is considered as a barrier; for example, any object on a pedestrian pathway that reduces the pedestrian corridor to less than 36" would be considered a barrier. In addition, features that change the surface height of the pedestrian pathway noticeably would also be recorded as a barrier. This kind of barrier may occur where tree roots push part of the sidewalk up, or where elevation discontinuities between sections are caused by inferior construction. Sewer grates or steam vents can also result in such situations, creating obstacles for mobility impaired individuals.

In addition, other features that were recorded that contained curb ramps. The characteristics of the ramps such as ramp type, slope, and presence/absence of landing at the top of the ramp were noted. These characteristics influence impedance analysis for mobility-impaired individuals, because they determine whether or not the ramps are ADA compliant. Due to the inherent limitation of the Network Analyst tool, this category of characteristics was recorded in the pedestrian network dataset as part of the road crossing features. Field checking for the JARC project consisted of two observation types: walking and viewing from a car. The field checking process was conducted twice in summer 2007 and summer 2011.

In pedestrian network analysis, the term impedance means the cost of movement. Distance will be used as the travel cost to estimate what pedestrians can reach within the network. A 1/4 mile has widely been considered as an accessible distance in previous pedestrian accessibility studies, which is equal to a 15-minute walk based on a normal walking rate (3 mile per hour). All line features within the pedestrian network were assigned an impedance value by calculation in advance.

In summary, Figure 3 illustrates the creation of Spokane Pedestrian Network. Spokane Pedestrian Network includes two critical components, Pedestrian Path Dataset and Barriers Dataset, both of which are initially created in JARC program. Spokane Pedestrian Network integrates these two parts and, in addition, impedance values are assigned to all line features in the dataset.

3.2.2 Missing Sidewalks Data Process

One key feature that was missing in the pedestrian network dataset before 2013 was values that code the block faces by sidewalk presence. This coding is critical. It is important to know the presence or absence of sidewalks along a road in the calculation of accessibility in the Network Analyst tool. This coding process was finished on March, 2013 according to the following steps.

a. Creating a Road Network without Alleys or Highways

The first step is to create a Spokane road feature class that does not include alleys or interstate highways. This feature class has a "key" attribute that has a unique identifier for

CHAPTER THREE METHODOLOGY

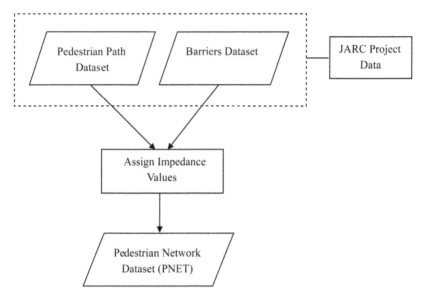

Figure 3: Spokane pedestrian network creation.
Source: Author.

block-to-block/intersection-to-intersection street segments. There were some KEY values missing, so we added them. The pedestrian network was based on an approximate 2005 road dataset and contained the keys for that year. The road feature class file "rd2013_k" was gained from the Spokane County GIS site with updates of the KEY field to match Pedestrian_Network2013.

The aim of creating a road coverage file is to create polygons that are defined by the road segments in order to include the information of the sidewalks on both sides in the road feature class. However, the "road2013" file contained alleys and highways that should not be included in the calculation of the block polygons. As a result, the first step was to create a road layer without alleys or highways. The types of the road segments were recorded in the "FTYPE" attribute. "Al" stands for an alley road, and "Hy" stands for the I-90 interstate highway. The "Select by attribute" tool was used to identify these alley and highway segments in this road feature class. A new selection is created with the input of "FTYPE='Al' OR FTYPE='Hy'".

All alley and highway segments were selected in the table, and the "switch selection" tool was used to selected all the other segments. The selected features are then exported as a new feature class, "rd2013_k_noalley." Similar to the "rd_2013" feature class, this feature class also contains an attribute named "key" that has a unique identifier for each block segment. In addition, the "key" value of the road segment is consistent with the "key" attribute of the parallel sidewalk segment in the "sw_4_f_2012" PNET feature class.

b. Create Block Polygons from Roads Layer

In the ArcInfo topology tool, polygons are built based on road line features and output the created coverage feature class. The GIS coverage file is created based on the road feature class (without alleys and the highway). The "Create coverage" tool can be used to create polygons that can be used to identify the sidewalks on each side of certain road segments. This step is processed by executing "Feature class To Coverage" tool. The input feature class is "road2013_noAlley," and the output feature class is "rd2013cov."

Then the next step is to execute the "Build Coverage" tool. The input feature class is the newly built coverage file "rd2013cov," and the feature type is chosen as "POLY."

The output polygon features are shown as follows:

Figure 4: Polygons divided by road network in the Spokane metropolitan area.
Source: Author.

c. Intersect Missing Sidewalk Feature Class with Block Polygons

The missing sidewalk feature class intersects with these block polygons and joins all available attributes. The result is that each of the missing sidewalk segments are within one polygon. We can determine for each of the roads segment which polygon is on its left or right side (since each sidewalk is on the left or right polygon for each street segment).

d. Calculate the Length of Missing Sidewalks

Multiple possible missing segments are summarized within each polygon and for each "key" value, and the results are recorded by executing "Summary Statistics."

For the new layer from the intersection between the missing sidewalk feature class and the block polygons, "Shape_Length" is selected as the Statistics Field, and "SUM" is selected as the summary statistics. The output table represents the shape length of each type of sidewalk (no sidewalk, sidewalk, road crossing, etc.) by each KEY value within each polygon.

e. Field Calculation

In the newly created summary table, a new key field that concatenates the POLY and KEY fields is created (one key for the left and one for the right side of each road segment).

Also, new key fields are created in the layer that contains all the missing sidewalk segments. Then, the road layer is joined with the missing sidewalk layer based on the new key sequentially for left the and right.

Next, the missing sidewalk segment layer is joined with the summary table twice by the left and right poly keys, and the total length of the left/right missing sidewalk is calculated in the "sum of the lengths" field. After this coding process, we can determine the presence or absence of the sidewalk and the length of missing links of sidewalks for each segment of the road network feature class based on each "Key" value.

3.2.3 Life-Needs Service Facilities Location Dataset

The detailed GIS-based data of Life-Needs Service Facilities Locations is classified by the North American Industrial Classification System (NAICS). NAICS is a standard used by federal statistical agencies to classify business establishments according to various types of economic activities for the purpose of collecting, analyzing, and publishing statistical data in the United States (United States Census Bureau, 2012). Each establishment is classified and coded to an industry based on its primary business activity. The categories are organized based on the definition and classification of Life-Needs Service Facilities discussed in Chapter 2.

a. Daily Food Providers

This category of Life-Needs Service Facilities generally contains one type of

3.2 Data Preparation Procedure

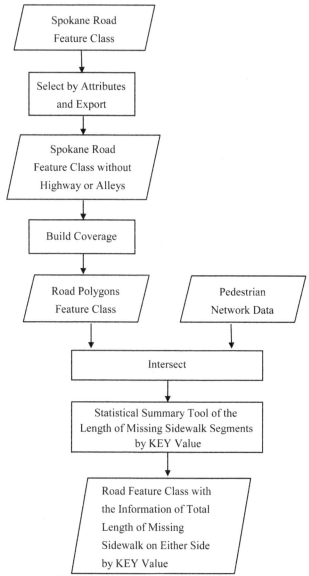

Figure 5: Missing sidewalks data process.
Source: Author.

establishment in NAICS: (445110) Supermarkets and Other Grocery Stores. This type of establishment comprises supermarkets and grocery stores that are primarily engaged in retailing a general line of food and are able to provide food services that satisfy the basic daily living needs of citizens. All Daily Food Providers within the study area are shown in Appendix A.

b. Social Gathering

This category includes one type of establishment in NAICS: (813110) Religious Organizations, which include churches, mosques, and other religious temples that administer organized religious activities. In addition, this category includes libraries and community centers, and their locations are acquired from the City of Spokane GIS Page. All Social Gathering facilities within the study area are shown in Appendix B.

c. Sports and Recreation Facilities

This category of Life-Needs Service Facilities generally contains two types of establishments in NAICS: (713940) Fitness and Recreational Sports Centers and (713990) All Other Amusement and Recreation Industries. (713940) Fitness and Recreational Sports Centers include establishments that provide fitness and recreational sports facilities for exercise and other sports activities. The category of (713990) All Other Amusement and Recreation Industries comprises all other establishments that provide recreational and amusement services. The Sports and Recreation facilities category includes fitness centers/gymnasiums, sports clubs, golf courses, bowling centers, swimming pools, recreation center, and other amusement and recreation facilities. All Sports and Recreation Facilities within the study area are shown in Appendix C.

d. Healthcare Facilities

This category contains three types of establishments in NAICS: (621) Ambulatory Health Care Services, (622) Hospitals, and (6231) Nursing Care Facilities. The first comprises a wide range of facilities in which health practitioners operate independent practices of general or specialized medicine, including dentists, optometrists, physical therapists, psychologists, etc. The second one comprises establishments providing diagnostic and medical treatment to patients with a wide variety of medical conditions. The last one comprises establishments that provide residential nursing care required by the residents. All Healthcare Facilities within the study area are shown in Appendix D.

e. Park Entrances

This category of Life-Needs Service Facilities generally includes parks. The raw data is gained from the Spokane County GIS website, which records the addresses of parks as a point feature class. In Network Analyst, the accessibility analysis is calculated based on the locations of geospatial points, which are usually gained from the addresses. Considering that parks in Spokane usually have multiple entrances for the surrounding neighborhoods that are different from the points that are created from addresses,

3.2 Data Preparation Procedure

calculation based on addresses would not be accurate in small-scale pedestrian accessibility analysis, since an accessible distance is defined as 1/4 mile, which is relatively small compared with the size of a park. As a result, a new point feature class with actual pedestrian entrances of the parks is needed for the destinations in later accessibility analysis. The recording of the entrances is processed manually based on Google Street View.

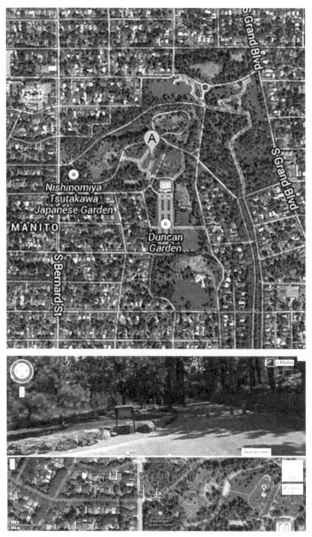

Figure 6: **Entrance identification by satellite map and street view image for Manito Park with Google StreetView.**

Source: Google Map.

CHAPTER THREE   METHODOLOGY

As an example, it's the address of Manito Park from the city GIS database is marked on Garland Boulevard, but residents may access this park from various entrances. Google Street View was used to identify the potential entrances to the park for pedestrians one by one.

Then, the entrances are recorded manually in a new point feature class in ArcMap Editor for future analysis.

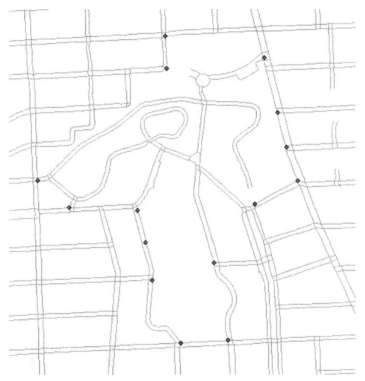

Figure 7: Edited point feature class of park entrances in ArcMap Editor.
Source: Author.

The spatial location of Life-Needs Service Facilities of these five categories are recorded separately in five point feature classes. Figure 8 shows a small-scale sample map of the point feature classes in ArcGIS. Different colors indicate Life-Needs Service Facilities in different categories.

52

3.2 Data Preparation Procedure

Figure 8: Life-Needs Service Facilities locations recorded as point feature classes.

Source: Author.

CHAPTER THREE   METHODOLOGY

3.2.4 Demographic Data

The first part of the demographic dataset contains Census Block Centroids in a point feature layer. The point features are created from a Census block polygon shapefile from the City of Spokane GIS page. The American Community Survey provides a Microsoft Access database shell that contains the 2010 Census summary file, including 2010 Vehicle Ownership data at the Census block group level. The FTP version of the 2010 Census SF1 data is released by the state as a series of files within a single compressed .zip file. Focus Census block groups were selected based on the vehicle ownership rate, which was calculated based on the data of total population and the number of vehicles for each Census block group from the 2010 American Community Survey data. Figure 9 presents an example of the vehicle ownership rate calculation. The Census block group with the GEOID 530630140013 has a total population of 1857. In addition, the residents in this Census block group own a total of 671 vehicle. Then the vehicle ownership rate of this Census block group is 671/1857 = 0.361335487, which is less than 0.5. Therefore, Census block group 530630140013 is considered as one of the focus Census block groups.

G	H	EQ	ER	ES	ET	EU	EV
GEOID	Geo Name	B25046	B25046	B25046	Veh icle	Total Popul	Vehicle per person
530630140013	Block Group 3, Census Tract 140.01, Spokane County, Washingto	671	392	279	0.6	1857	0.361335487

Figure 9: Vehicle ownership rate calculation at the level of Census group block.
Source: American Community Survey 2010.

All the Census blocks within the selected Census block groups are considered as the focus Census blocks. Figure 10 presents a zoomed-in area as an example of the selection results. The dark red lines represent the boundaries of the Census block groups with low vehicle ownership rate. The light red polygons represent the Census blocks that locate within these Census block groups and thus are defined as focus Census blocks. In addition, for these focus Census block, 2010 Census block population data is used as the main population source in the analysis. The centroids of these focus Census block are recorded as the point feature class "Focus_BlockCentroids."

3.2 Data Preparation Procedure

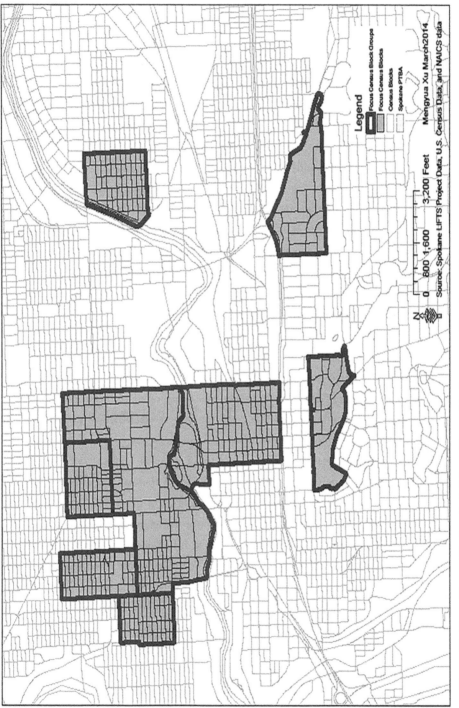

Figure 10: Focus Census blocks selected from Census block groups with a low vehicle ownership rate.
Source: American Community Survey 2010.

CHAPTER THREE METHODOLOGY

3.3   Data Analysis Procedure

The main aim of this data analysis is to identify the missing sidewalk segments that affect the level of accessibility to Life-Needs Service Facilities in Spokane PTBA and thus prioritize the potential improvement based on both equity and efficiency criteria. Therefore, the first step is to identify all missing sidewalk segments that may be important for the level of accessibility. The basic identification method is to calculate the potential routes from Census block centroids to the closest Life-Needs Service Facility of each category in an ideal pedestrian network and select the missing sidewalk segments on these routes. In other words, if one missing sidewalk segment located on the ideal routes from Census block centroids to Life-Needs Service Facilities, this segment is considered as important for accessibility improvement. The routes from each Census block centroid to closest Life-Needs Service Facility are calculated separately by the five categories of Life-Needs Service Facilities. In addition, it is possible that some missing sidewalk segments may be selected more than once, which weights the importance for these missing sidewalk segments.

The next step is to determine the importance of each missing sidewalk segment for prioritization with quantitative measures. As discussed in Chapter 2, the prioritization of potential fixing of the missing sidewalk segments was determined from the perspectives of Efficiency and Equity. The final importance measure was a combination of the two. The measure of Efficiency Importance of each missing sidewalk segment was based on the its length (which is related to the cost of fixing) and the number of residents who would have access to nearby Life-Needs Service Facilities destinations if this missing sidewalk segment were fixed (the population data of the number of people is acquired from the Census block). To be more specific, if there are more people using the routes that include one specific missing sidewalk segment to access the closest Life-Needs Service Facilities in the five categories, then this missing sidewalk segment is more important and thus has a higher priority. If one specific missing sidewalk segment is relatively short, then it costs less to fix it, which also gives it a higher prioritization for fixing due to economic efficiency as well.

In this thesis, the focus group is defined as the mobility disadvantaged group who has less access to a private car and thus heavily relied on pedestrian access to services. Therefore the focus Census blocks are selected based on the calculation of vehicle ownership rate. 2010 Vehicle ownership data from American Community Survey is used as the source for the calculation of 2010 Vehicle Ownership rate at the Census block group level.

To conclude, the Efficiency Importance of each missing sidewalk segments was based on the its length (which related to the cost of fixing) and the number of residents (the population data of number of people is acquired from the Census block) who would have access to nearby Life-Needs Service Facilities destinations if this missing sidewalk segment is fixed. Similarly, the Equity Importance of one missing sidewalk segment is defined as the accumulated numbers of residents may be served by this segment if it is fixed based on potential routes from focus Census blocks to the nearest destinations for each categories of Life-Needs Service Facilities, divided by the length of this segment.

Figure 11 illustrates the framework of how to operate the data analysis procedure.

3.3.1 Complete Pedestrian Network Creation

For the needs of future analysis and prioritization, it is necessary to first create an ideal pedestrian network for further route calculation. The complete pedestrian network represents the ideal situation for any pedestrians and shows the maximal extent of the potential pedestrian accessibility. In this network, all sidewalks are considered complete and ADA accessible to accommodate a walking rate of 3 miles per hour, the same as in the PNET. The complete pedestrian network was created from an original pedestrian network dataset, and all line features in the complete pedestrian network are identical to the original ones.

3.3.2 Routes Calculation

Routes in this study are defined as the Origin-Destination (OD) Pairs that are calculated as the shortest routes from Census block centroids to various Life-Needs Service Facilities locations. They are calculated in the ArcGIS Network Analyst toolset, and the cut-off travel time used in the present study was set to 1/4 mile (400 meters), the distance required for 15 minutes of walking at an average speed of 3 miles per hour. The destinations beyond the cutoff travel time in the particular pedestrian network would not be considered as accessible.

The aim of the calculation of the route is to identify the important pedestrian pathways in the network. The missing sidewalk segments on the routes from Census block centroids (where the populations are located) to the Life-Needs Service Facilities locations (where the services are located) are considered as important. Pedestrian routes were calculated for every Census block centroid and the closest instance of each Life-Needs Service Facilities category in the Complete Pedestrian Network. The calculation is facilitated by the Network

CHAPTER THREE   METHODOLOGY

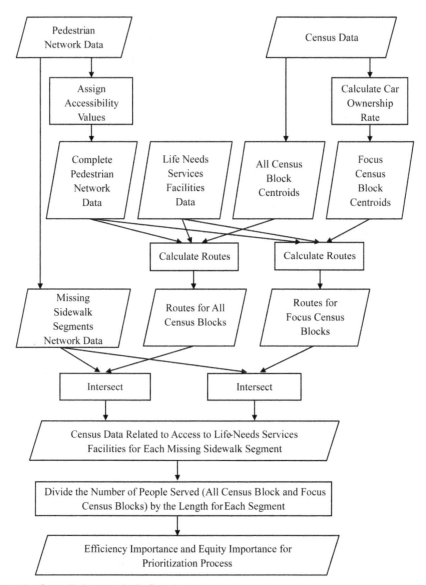

Figure 11: Overall data analysis flowchart.
Source: Author.

Analyst tool in ArcGIS 10.2. The detailed steps of route calculations are as follows.

The first step is to load Census block centroids and destinations. In the Network Analysis toolbar, click New Closest Facility. Then, right-click Facilities and choose Load Locations.

In the Load Location Dialogue box, choose the point feature class of the first category of Life-Needs Service Destinations: "Grocery_supermarkets" from the Load From drop-down list. Choose the "DestID" Field as the name for the convenience of future data analysis. In addition, the cutoff-length of the analysis is set as 1320 feet (which is 1/4 mile), and the search tolerance is set as 100 feet. The grocery stores are added and listed in the Network Analyst window.

Similarly, right-click Incidents and add the "Focus_BlockCentroids" feature class from the drop-down list. Choose the "GEOID10" field as the name for identification needs of future data analysis and set the search tolerance as 100 feet. In the Layer Properties window, set the Default value of "Name" as GEOID10, which is a unique value that can be used to identify the Census Blocks.

After the Facilities and Incidents are both loaded, click the "Solve" button on the Network Analyst window, where the routes are generated and appear. Export the calculated routes and save as a new line feature class, "Route_FocusBlock_Groc."

The next step is to transfer the population data from the Census blocks to the route line features. First, a new field "Block_Pop" is created in the "Route_FocusBlock_Groc" route line feature class. The type of field is set as "Long Integer," since this field records the population numbers of the specific Census block related to each route.

Then, use the "Join" tool to append the data from the "Focus_BlockCentroids" feature class to the routes. During the process of route calculation, the GEOID of the Census block for each corresponding route is automatically recorded as one field of the "Route_FocusBlock_Groc" feature class, which can be used as the field that the joint is based on.

The overall population information is recorded in the "POP10" field in the Census block centroids feature class. After the joint, use the field calculator to calculate the "Block_Pop" field of the route feature class. After the calculation, it is needed to remove the join. Then, repeat the same process of route calculation and field calculation for each category of the Life-Needs Service Facilities.

Similarly, load the "BlockCentroid_within_PTBA" feature class, which includes all Census block centroids within the study area as Incidents and repeat the whole process. At this point, all potential routes that represent the Origin-Destination pair from Census block centroids to multiple Life-Needs Service Facilities locations are created and recorded within categories based on the type of Census block (focus Census block or not) and Life-Needs Service Facilities. As there are five categories of Life-Needs Service Facilities, by relating each category with focus Census blocks as well all Census blocks, ten groups of routes are created.

Figures 12-14 show some samples of the calculated routes and Census Block 1036 is selected as the sample Census block. Figure 11 and Figure present the shortest routes from Census Block 1036 to the closest Daily Food Provider and Social Gathering Facility, respectively. And there is no Healthcare Facility or Park Entrance within walking distance (1/4 mile) from Census Block 1036. Therefore, no route is created in these two categories, as shown in Figure 13.

3.3.3 Intersecting Missing Sidewalk Layer with the Routes

The aim of this step is to relate the potential route with the missing sidewalk and transfer the information recorded within the routes layer to the missing sidewalk layer in order to determine the importance of each missing sidewalk segments for prioritization. First of all, it is necessary to create a line feature class for the missing sidewalk segments. In the Pedestrian Network feature class, select by attribute of PDPTHTYPE with value of 0.

Output the selected data to create a layer with all missing sidewalks with the name of "All_Missing_SW." In the attribute table of the "All_Missing_SW" layer, create a new attribute with the name of "SW_ID" for further joint and identification. Then, use the Field Calculator to set "SW_ID" to ObjectID.

Then, use the Intersect tool to intersect each route feature class with the "All_Missing_SW" feature one by one. As mentioned before, there are ten groups of various routes in total. Taking the routes from focus Census block centroids to grocery stores as an example, in order to generate the overlapped line features from the routes data and the missing sidewalk segments, click the "Input Features" drop-down list in the "Intersect" tool window and choose the "All_Missing-SW" and "Route_FocusBlock_Groc" layers. The output data is a line feature class that keeps the overlapped line segments from both of the feature classes and combines both attributes.

Obviously, some of the missing sidewalk segments may be selected by multiple routes and thus related to different Census blocks. In order to determine the overall population number that each missing sidewalk serves for each Life-Needs Service Facilities category, it is necessary to first summarize the population number, and the "SW_ID" field can be used to identify the segments, since each segment has a unique "SW_ID" value. As a result, the next step is to use the "Summarize" tool to create a new table to sum and record the population data for each missing sidewalk segment. In the "Summarize" tool window, select the "SW_ID" field in the drop-down list to summarize, because it is the

3.3 Data Analysis Procedure

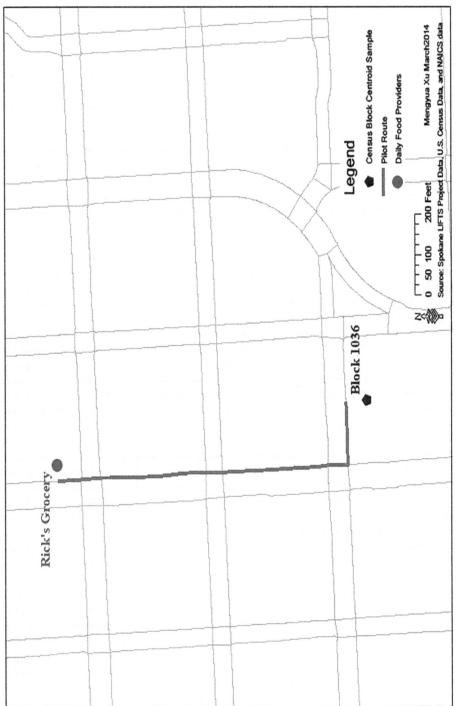

Figure 12: Route from Census Block 1036 to the closest Daily Food Provider.
Source: Author.

CHAPTER THREE   METHODOLOGY

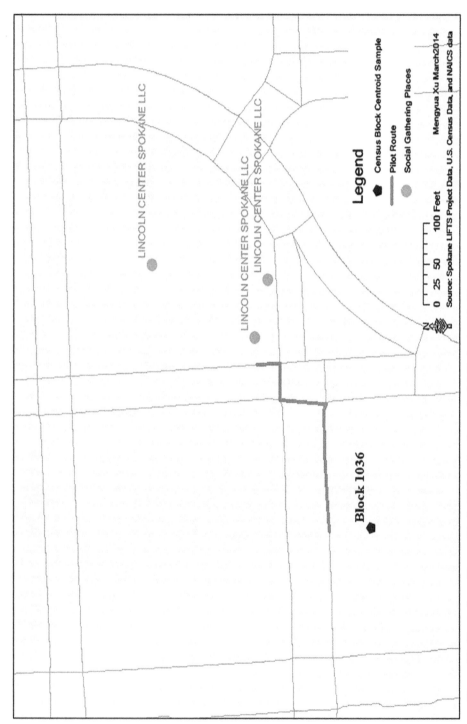

Figure 13: Route from Census Block 1036 to the closest Social Gathering Facility.
Source: Author.

3.3 Data Analysis Procedure

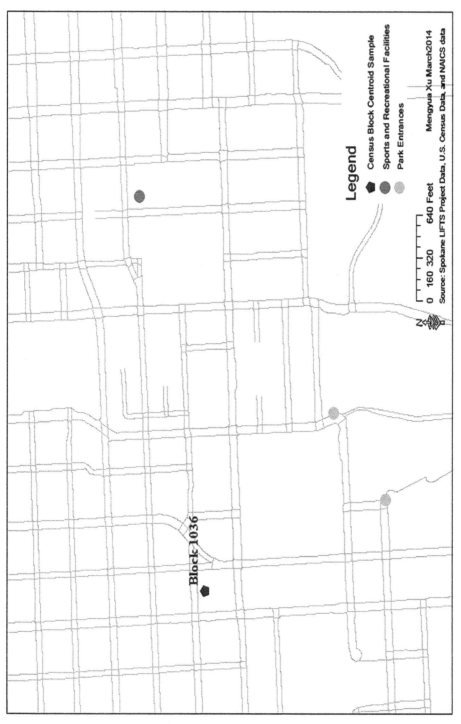

Figure 14: No Healthcare Facility or Park Entrance from Census block 1036 within walking distance.
Source: Author.

CHAPTER THREE METHODOLOGY

unique ID for each segment. Choose "Sum" statistics for the "Block_Pop" field. The output summary table calculates the accumulated population number for each "SW_ID" and records the results in the "Sum_Block_Pop" field.

Then, join the Missing SW layer with the summary table by the attribute of "SW_ID." As mentioned before, the "SW_ID" field acts as a unique value to identify each segment. So, choose "SW_ID" as the field that the join is based on in the drop-down list. Similarly, choose "SW_ID" as the field in the summary table that the join is based on.

After the join, create a new field "Pop_FBlock_Groc" in the Missing Sidewalk line feature class to record the accumulated number of population. Use the Field Calculator tool to make the "Pop_FBlock_Groc" field equal to the "Sum_Block_Pop" field in the summary table.

Then, repeat the Intersect process for all the groups of routes based on different combinations between Life-Needs Service facilities destinations categories and all Census blocks/focus Census blocks. Next, create the corresponding fields and join table one after the other. Use the field calculator to record the population data from Census block in the missing sidewalk feature class by categories of Life-Needs Service Facilities.

3.3.4 Importance Calculation

At this point, for each missing sidewalk segment, the population number it serves for the connection between a Census block and service destination is recorded by the categories of Life-Needs Service Facilities and types of Census blocks (either focus or non-focus Census blocks).

Since different categories of the Life-Needs Service Facilities are considered as equally important in the present data analysis process, the next step is to sum the population number served for the routes to each category of Life-Needs Services Facilities. Create new long integral Fields "Pop_Sum_AllBlock" and "Pop_Sum_FBlock." Use the Field Calculator to sum the population number by categories for all Census blocks and focus Census blocks. This summarization contains the population numbers from calculation results from all categories of Life-Needs Services Facilities.

As mentioned before, the Efficiency Importance Measure of each missing sidewalk segment is defined as the accumulated numbers of residents that may be served by this segment if it is fixed based on potential routes from all Census blocks to the nearest destinations for each category of Life-Needs Service Facilities, divided by the length of this segment.

The formula of **Efficiency Importance Measure** calculation can be expressed as

follows:
$$I_{eff} = (\sum P_{groc} + \sum P_{gat} + \sum P_{spo} + \sum P_{hea} + \sum P_{pa}) / L_n$$
where

I_{eff} is the Efficiency Importance Measure;

L_n is the shape length of the missing sidewalk segment;

$\sum P_{groc}$ is the total number of people that this particular segment serves for the routes linking the Census block centroid to the closest grocery store;

$\sum P_{groc} = P_{groc}1 + P_{groc}2 + P_{groc}3 + \cdots + P_{groc}n$ ($P_{groc}n$ is the population size of one Census block if this particular segment is on the route linking this Census block centroid to the closest grocery store);

Similarly,

$\sum P_{gat} = P_{gat}1 + P_{gat}2 + P_{gat}3 + \ldots + P_{gat}n$ ($P_{gat}n$ is the population size of one Census block if this particular segment is on the route linking this Census block centroid to the closest gathering space);

$\sum P_{spo} = P_{spo}1 + P_{spo}2 + P_{spo}3 + \ldots + P_{spo}n$ ($P_{spo}n$ is the population size of one Census block if this particular segment is on the route linking this Census block centroid to the closest sports and recreation facility);

$\sum P_{hea} = P_{hea}1 + P_{hea}2 + P_{hea}3 + \ldots + P_{hea}n$ ($P_{hea}n$ is the population size of one Census block if this particular segment is on the route linking this Census block centroid to the closest healthcare and social assistance facility);

$\sum P_{pa} = P_{pa}1 + P_{pa}2 + P_{pa}3 + \ldots + P_{pa}n$ ($P_{pa}n$ is the population size of one Census block if this particular segment is on the route linking this Census block centroid to the closest park entrance);

At the same time, the **Equity Importance Measure** of each missing sidewalk segment is defined as the accumulated numbers of residents may be served by this segment if it is fixed based on potential routes from focus Census blocks to the nearest destinations for each category of Life-Needs Service Facilities, divided by the length of this segment.

The formula of Equity Importance Measure calculation can be expressed as follows:
$$I_{eq} = (\sum FP_{groc} + \sum FP_{gat} + \sum FP_{spo} + \sum FP_{hea} + \sum FP_{pa}) / L_n$$
where

I_{eq} is the Equity Importance Measure;

L_n is the shape length of the missing sidewalk segment;

$\sum FP_{groc}$ is the total number of people that this particular segment serves for the routes linking the focus Census block centroid to the closest grocery store;

$\sum FP_{groc} = P_{groc}1 + P_{groc}2 + P_{groc}3 + \ldots + P_{groc}n$ ($P_{groc}n$ is the population size of one

CHAPTER THREE METHODOLOGY

focus Census block if this particular segment is on the route linking this Census block centroid to the closest grocery store);

$\sum FP_{gat} = P_{gat}1 + P_{gat}2 + P_{gat}3 + \ldots + P_{gat}n$ ($P_{gat}n$ is the population size of one focus Census block if this particular segment is on the route linking this Census block centroid to the closest gathering space);

$\sum FP_{spo} = P_{spo}1 + P_{spo}2 + P_{spo}3 + \ldots + P_{spo}n$ ($P_{spo}n$ is the population size of one focus Census block if this particular segment is on the route linking this Census block centroid to the closest sports and recreation facility);

$\sum FP_{hea} = P_{hea}1 + P_{hea}2 + P_{hea}3 + \ldots + P_{hea}n$ ($P_{hea}n$ is the population size of one focus Census block if this particular segment is on the route linking this Census block centroid to the closest healthcare and social assistance facility);

$\sum FP_{pa} = P_{pa}1 + P_{pa}2 + P_{pa}3 + \ldots + P_{pa}n$ ($P_{pa}n$ is the population size of one focus Census block if this particular segment is on the route linking this Census block centroid to the closest park entrance);

3.3.5 Double-Checking Missing Sidewalk Segments on Both Sides of the Roads

Due to the limitations of the calculation methodology of the Network Analyst toolset, the route calculation may pick up the pedestrian sideways with only one side containing missing sidewalk, while the other side of the road is accessible. In this situation, the pedestrian pathway may still be accessible in general, and the selected missing sidewalk segments should not have high prioritization for improvement. As a result, it is essential to examine the condition of the pedestrian pathway on both sides of this particular road segment for every missing sidewalk segment selected with an overall importance in the last step. The total length of missing sidewalk has been calculated and recorded by KEY value in the road feature class in previous analysis. The missing sidewalk layer and the road layer cannot be related by the "Key" value, so it can be easily examined. If the selected segment has no missing sidewalk on the other side of the road, then it would be not recorded and calculated.

3.3.6 Overall Prioritization Determination for Missing Sidewalk Segments

Again, the Efficiency Importance of one missing sidewalk segment is defined as the accumulated numbers of residents that may be served by this segment if it is fixed based

3.3 Data Analysis Procedure

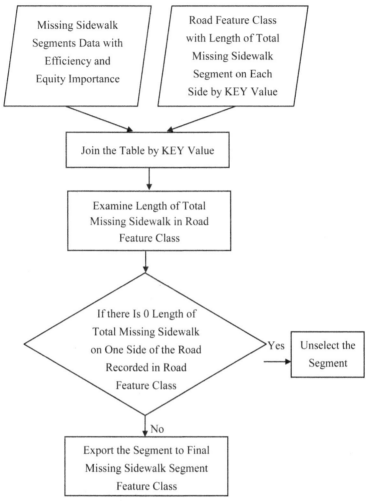

Figure 15: Double-checking missing sidewalk segments on both sides of the roads.
Source: Author.

on potential routes from all Census blocks to the nearest destinations for each category of Life-Needs Service Facilities, divided by the length of this segment. Similarly, the Equity Importance of one missing sidewalk segment is defined as the accumulated numbers of residents that may be served by this segment if it is fixed based on potential routes from focus Census blocks to the nearest destinations for each category of Life-Needs Service Facilities, divided by the length of this segment. In other words, the Equity Importance values are only applied to the missing sidewalk segments that are relate to the routes

CHAPTER THREE METHODOLOGY

connecting focus Census blocks with Life-Needs Service Facilities.

As illustrated in the previous chapter, a focus group is defined as a mobility-disadvantaged group that has less access to a private car and thus heavily relies on pedestrian access to services. Therefore, the focus Census blocks are selected based on the calculation of vehicle ownership rate.

The final prioritization decision should be based on both Equity Importance and Efficiency Importance. A link with high equity importance does not necessarily have an Efficiency Importance, and vice versa. The overall prioritization determination for missing sidewalk segments is based on the combination of both Efficiency Importance Measure and Equity Importance Measure as shown in Figure 16. The equation is:

Overall Importance = Efficiency Importance + Equity Importance * r

where

r = ratio of the emphasis between Efficiency and Equity

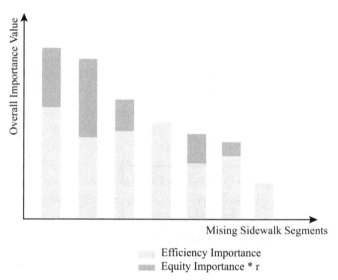

Figure 16: Prioritization of missing sidewalk segments.
Source: Author.

Figure 16 illustrates a mock result of prioritization based on Efficiency Importance and Equity Importance. Each column represents a combined Overall Importance Value of one missing sidewalk segment, listed from high to low in sequence of the value. Blue and green stripes in the columns represent the values of Efficiency Importance and Equity Importance for this missing sidewalk segment respectively.

The equation provides researchers some flexibility to adjust the r value based on different circumstances in order to better accommodate various research focuses. For example, the formula can place higher weight on Equity Importance by setting higher r value (2 instead of 0.5, for instance). In this situation, missing sidewalk segments with relatively high Equity Importance are given higher weights in Overall Importance calculation. When r value equals 0, Overall Importance equals to the Efficiency Importance, which means no additional weight is placed on focus Census blocks.

CHAPTER FOUR   RESULTS

The aim of the book is to develop a GIS-based model that can measure pedestrian accessibility and prioritize the potential improvement based on the criteria of equity and efficiency. This chapter illustrates the results of applying the methodology to the study area, the Spokane PTBA. This chapter first presents the vehicle ownership rate calculated at the Census block group level and the focus Census blocks that are defined by the relative low vehicle ownership rate. Secondly, this chapter presents the results of all potential routes connecting all Census block centroids (and focus Census block centroids) to the closest Life-Needs Service Facility in each category, as shown in Figure 19 - Figure 28. Thirdly, it presents results of the missing sidewalk segment selection based on the routes, as shown in Figure 31 - Figure 40. Fourthly, this chapter presents the results of importance calculation for the missing sidewalk segment, including Equity Importance and Efficiency Importance as well as Overall Importance. It also represents various final results scenarios based on different weights between Equity Importance and Efficiency Importance as shown in Figure 42 - Figure 46.

4.1   Census Block Centroids Locations

The Census block data was gained from U.S. Census data, and the following figure shows the spatial distribution of the Census block centroids within the Spokane PTBA.

Again, the selection of focus Census block groups is based on vehicle ownership rate. The following figure illustrates the locations of the Census block centroids of the selected ones. The rate of vehicle ownership is calculated based on the data of total population and the number of vehicles for each Census block from the 2010 American Community Survey data. The Census block groups with a rate of vehicle ownership lower than 0.50 vehicles per person was considered as the target neighborhoods. All the focus Census block groups

4.1 Census Block Centroids Locations

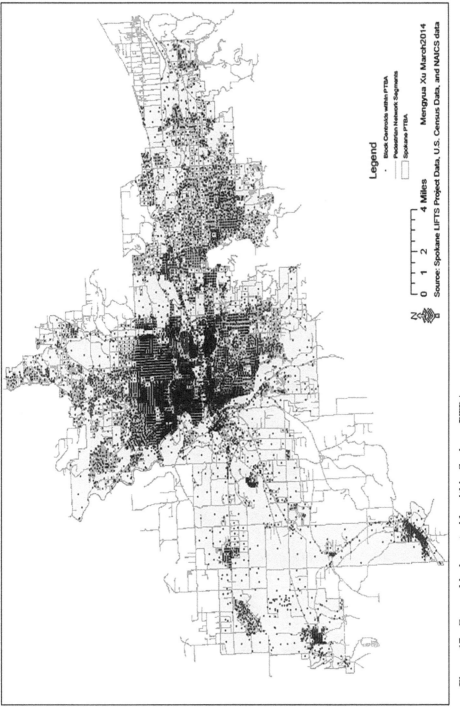

Figure 17: Census block centroids within Spokane PTBA.

Source: Author.

CHAPTER FOUR RESULTS

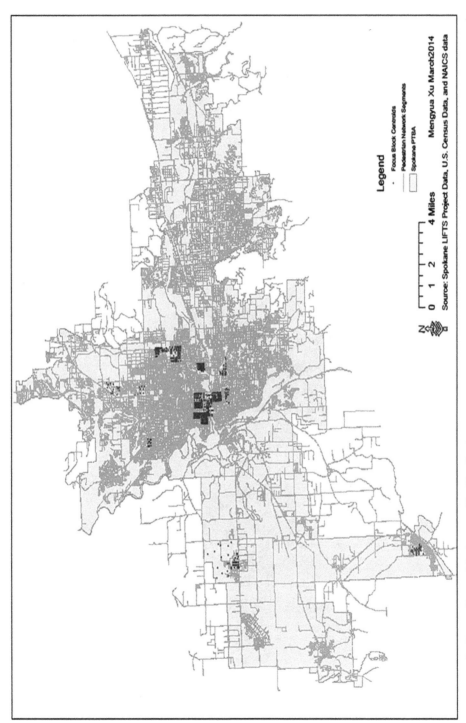

Figure 18: Focus Census block centroids distribution.

Source: Author.

identified by vehicle ownership rate are listed in Table 1. All the Census blocks within the selected Census block groups with low vehicle ownership rate are considered as the focus Census blocks. By this standard, 698 out of a total of 11537 Census blocks were identified as focus Census blocks in the study area. The population data of these focus Census blocks is illustrated in Appendix E.

Table 1: Focus Census block groups identified by vehicle ownership rate.

	D	E	F	G	H	EQ	ER	ES	ET	EU	EV
1	Itter#	SeqNum	LOGRECNO	GEOID	Geo Name	B25046	B25046	B25046	Vehicle	Total Popul	Vehicle per person
2	000	0097	0010032	530630140013	Block Group 3, Census Tract 140.01, Spokane County, Washingto	671	392	279	0.6	1857	0.361335487
3	000	0097	0010030	530630140011	Block Group 1, Census Tract 140.01, Spokane County, Washingto	269	.	269		1117	0.240823635
4	000	0097	0009932	530630111022	Block Group 2, Census Tract 111.02, Spokane County, Washingto	811	153	658	0.2	1707	0.475102519
5	000	0097	0009927	530630111011	Block Group 1, Census Tract 111.01, Spokane County, Washingto	453	99	354	0.2	1341	0.337807606
6	000	0097	0009899	530630104011	Block Group 1, Census Tract 104.01, Spokane County, Washingto	1082	676	406	0.6	3629	0.298153761
7	000	0097	0009831	530630035002	Block Group 2, Census Tract 35, Spokane County, Washington	157	31	126	0.2	784	0.200255102
8	000	0097	0009827	530630032002	Block Group 2, Census Tract 32, Spokane County, Washington	321	101	220	0.3	802	0.400249377
9	000	0097	0009822	530630031001	Block Group 1, Census Tract 31, Spokane County, Washington	741	333	408	0.4	1804	0.41075388
10	000	0097	0009815	530630026003	Block Group 3, Census Tract 26, Spokane County, Washington	550	344	206	0.6	1305	0.421455939
11	000	0097	0009806	530630024002	Block Group 2, Census Tract 24, Spokane County, Washington	458	115	343	0.3	1041	0.439961575
12	000	0097	0009805	530630024001	Block Group 1, Census Tract 24, Spokane County, Washington	396	143	253	0.4	1646	0.240583232
13	000	0097	0009801	530630023001	Block Group 1, Census Tract 23, Spokane County, Washington	686	328	358	0.5	1411	0.486180014
14	000	0097	0009797	530630020004	Block Group 4, Census Tract 20, Spokane County, Washington	599	315	284	0.5	1205	0.497095436
15	000	0097	0009787	530630016002	Block Group 2, Census Tract 16, Spokane County, Washington	385	168	217	0.4	977	0.39406346
16	000	0097	0009786	530630016001	Block Group 1, Census Tract 16, Spokane County, Washington	561	291	270	0.5	1369	0.409788167
17	000	0097	0009757	530630009001	Block Group 1, Census Tract 9, Spokane County, Washington	451	275	176	0.6	962	0.468814969
18	000	0097	0009733	530630002003	Block Group 3, Census Tract 2, Spokane County, Washington	299	236	63	0.8	827	0.361547763

Source: American Community Survey 2010.

4.2 Route Calculation

The results of route calculation describe the potential pedestrian access from Census block centroids to the closest destinations of each category of Life-Needs Service Facilities (if there are any within walking distance, which is defined as 1/4 mile in this research).

CHAPTER FOUR RESULTS

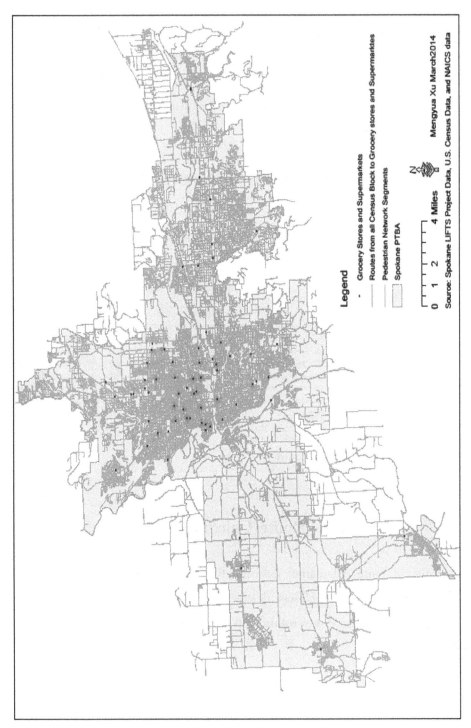

Figure 19: Routes from all Census block centroids to the closest Daily Food Provider within walking distance.
Source: Author.

4.2 Route Calculation

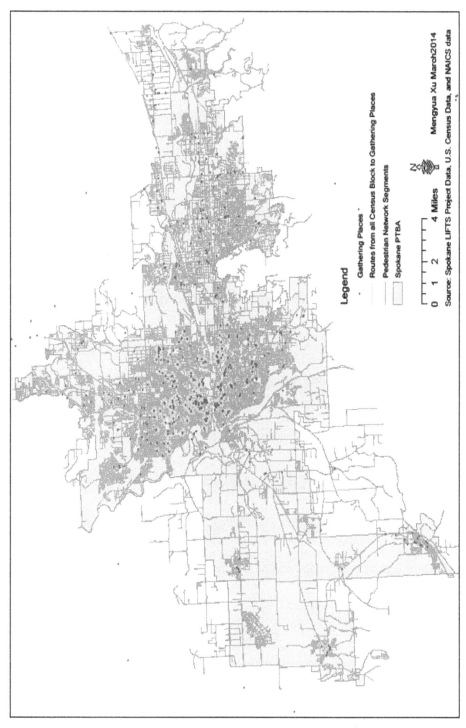

Figure 20: Routes from all Census block centroids to the closest Social Gathering Place within walking distance.
Source: Author.

CHAPTER FOUR   RESULTS

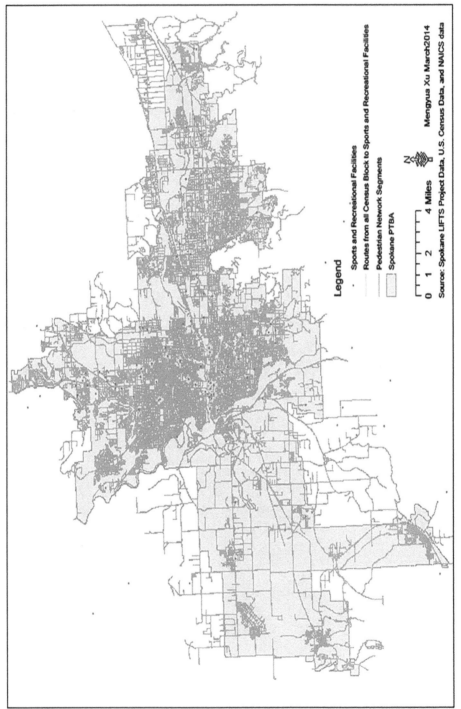

Figure 21: Routes from all Census block centroids to the closest Sports and Recreation Facility within walking distance.

Source: Author.

4.2 Route Calculation

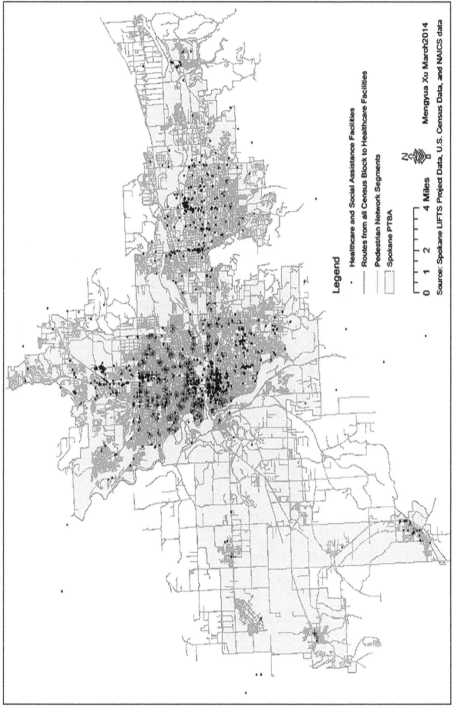

Figure 22: Routes from all Census block centroids to the closest Healthcare Facility within walking distance.

Source: Author.

CHAPTER FOUR RESULTS

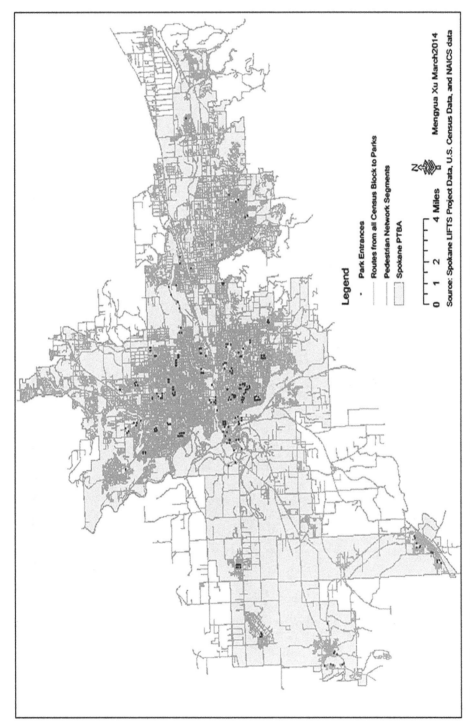

Figure 23: Routes from all Census block centroids to the closest Park Entrance within walking distance.

Source: Author.

4.2 Route Calculation

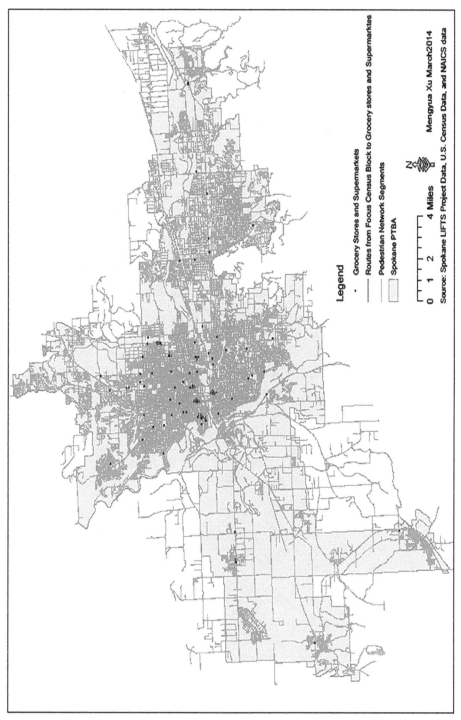

Figure 24: Routes from focus Census block centroids to the closest Daily Food Provider within walking distance.
Source: Author.

CHAPTER FOUR   RESULTS

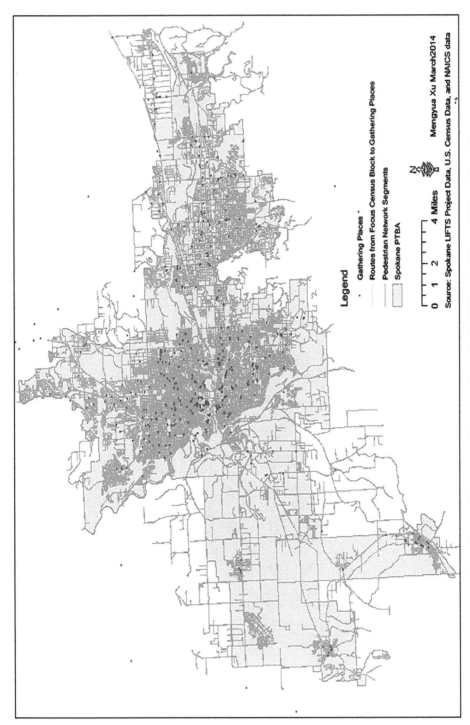

Figure 25: Routes from focus Census block centroids to the closest Social Gathering Place within walking distance.

Source: Author.

4.2 Route Calculation

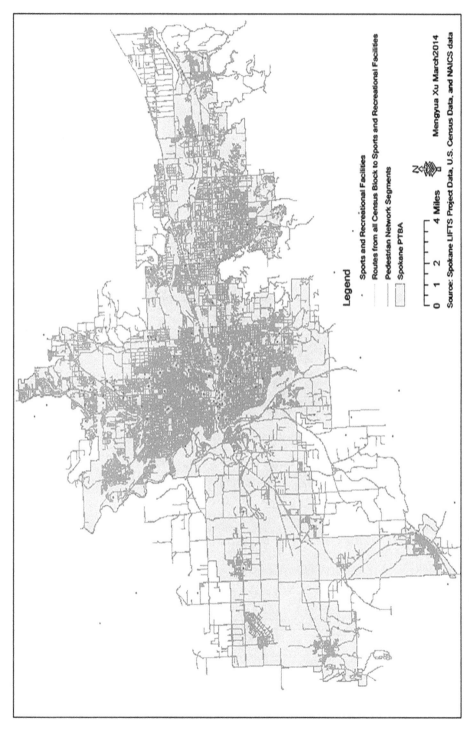

Figure 26: Routes from focus Census block centroids to the closest Sports and Recreation Facility within walking distance.
Source: Author.

CHAPTER FOUR   RESULTS

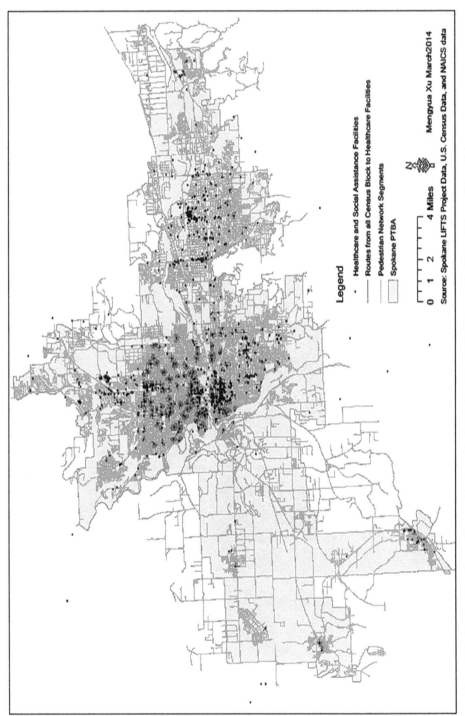

Figure 27: Routes from focus Census block centroids to the closest Healthcare Facility within walking distance.
Source: Author.

4.2 Route Calculation

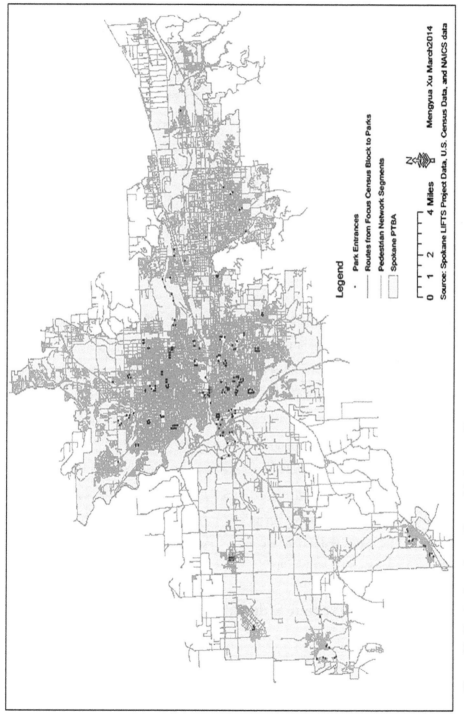

Figure 28: Routes from focus Census block centroids to the closest park entrance within walking distance.

Source: Author.

CHAPTER FOUR   RESULTS

From the results of route calculation, 3938 out of a total of 11537 (34.13%) Census blocks have access to at least one of the service destinations in the category of Daily Food Providers within walking distance (1/4 mile). 2509 out of a total of 11537 (21.75%) Census blocks have access to at least one of the service destinations in the category of Social Gathering Places. 448 out of a total of 11537 (3.88%) Census blocks have access to at least one of the service destinations in the category of Sports and Recreation Facilities. 2828 out of a total of 11537 (24.51%) Census blocks have access to at least one of the service destinations in the category of Healthcare Facilities. 903 out of a total of 11537 (7.83%) Census blocks have access to at least one of the service destinations in the category of Parks Entrances.

114 out of a total of 698 (14.56%) focus Census blocks have access to at least one of the service destinations in the category of Daily Food Providers. 320 out of a total of 698 (45.85%) focus Census blocks have access to at least one of the service destinations in the category of Social Gathering Places. 76 out of a total of 698 (10.89%) focus Census blocks have access to at least one of the service destinations in the category of Sports and Recreation Facilities. 314 out of a total of 698 (44.99%) focus Census blocks have access to at least one of the service destinations in the category of Healthcare Facilities. 75 out of a total of 698 (10.74%) focus Census blocks have access to at least one of the service destinations in the category of Parks Entrances.

Only 4 Census blocks in the study area have pedestrian access to at least one of the Life-Needs Service Facility in all five categories within walking distance, all of which locates in Airway Heights as shown in Figure 29. In addition, 7599 out of a total of 11537 Census block do not have pedestrian access to any Life-Needs Service Facility within walking distance, as shown in Figure 30.

4.3   Intersected Sidewalk from the Routes

The next step is to intersect the sidewalk network with the calculated routes in order to extract the missing sidewalk segments that are potentially important. The calculation was conducted for both focus Census blocks and all Census blocks. The following figures show the results:

4.3 Intersected Sidewalk from the Routes

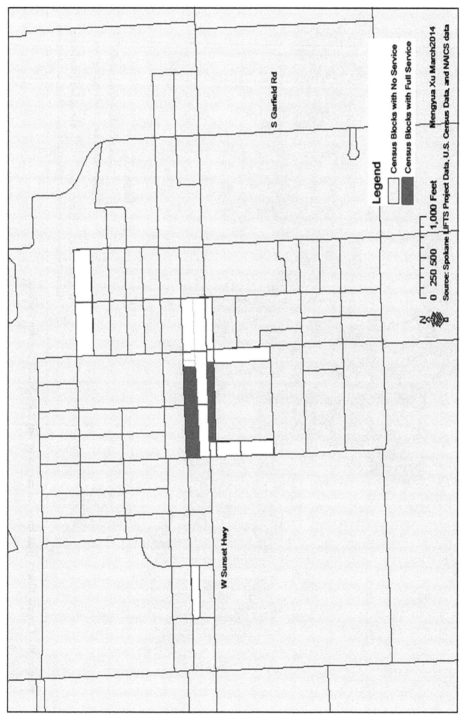

Figure 29: Census Blocks that have access to all five categories of **Life-Needs Service Facility**.
Source: Author.

CHAPTER FOUR   RESULTS

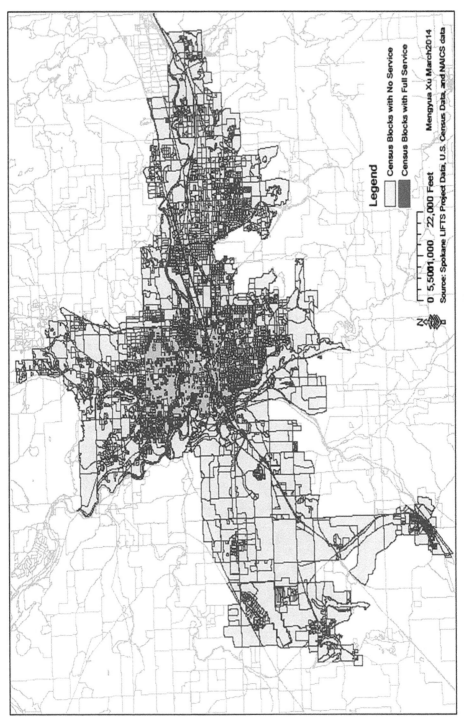

Figure 30: Census Blocks that have no access to any Life-Needs Service Facility.
Source: Author.

4.3 Intersected Sidewalk from the Routes

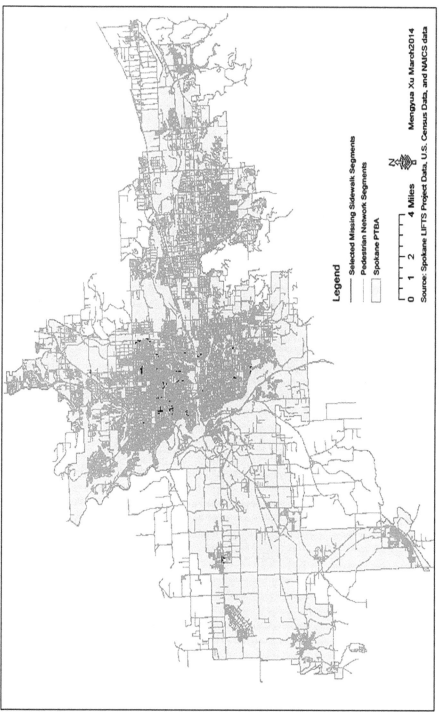

Figure 31: Missing sidewalk segments selected from routes linking all Census block centroids to the closest daily food provider within walking distance.

Source: Author.

CHAPTER FOUR RESULTS

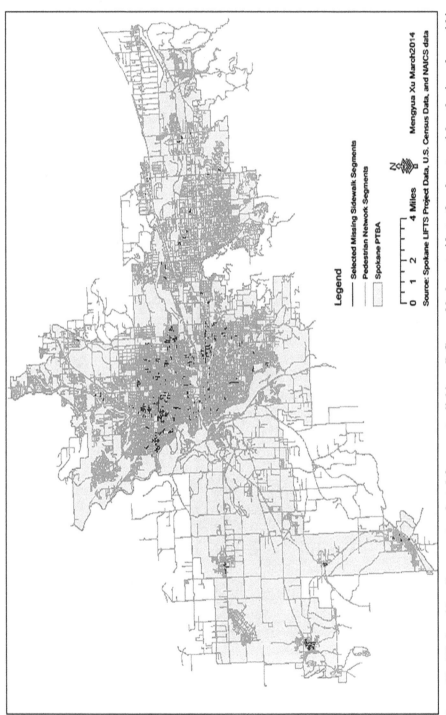

Figure 32: Missing sidewalk segments selected from routes linking all Census block centroids to the closest social gathering place within walking distance.

Source: Author.

4.3 Intersected Sidewalk from the Routes

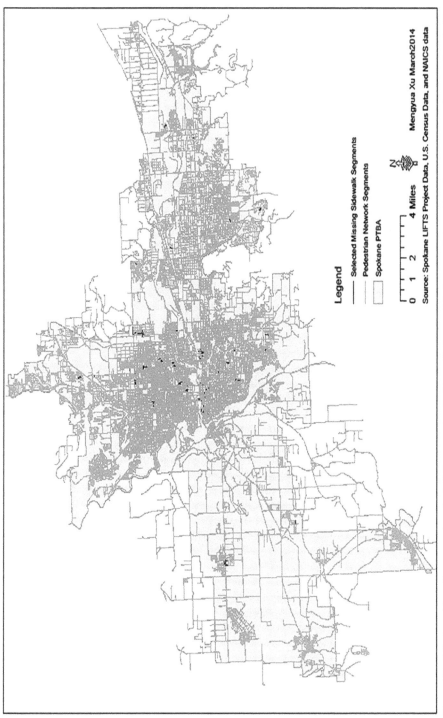

Figure 33: Missing sidewalk segments selected from routes linking all Census block centroids to the closest sport and recreation facility within walking distance.

Source: Author.

CHAPTER FOUR   RESULTS

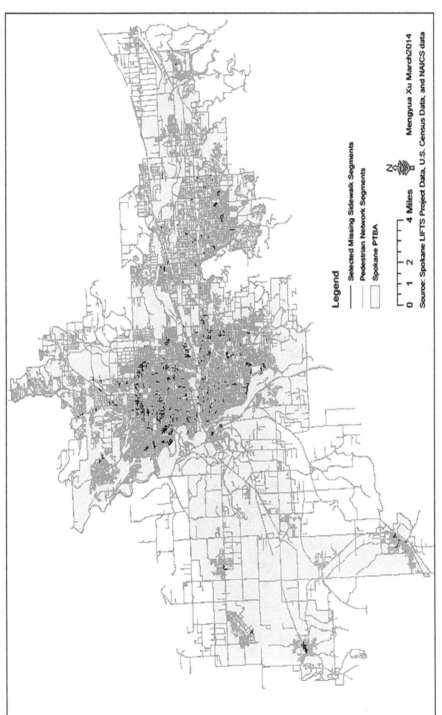

Figure 34: Missing sidewalk segments selected from routes linking all Census block centroids to the closest healthcare facility within walking distance.

Source: Author.

4.3 Intersected Sidewalk from the Routes

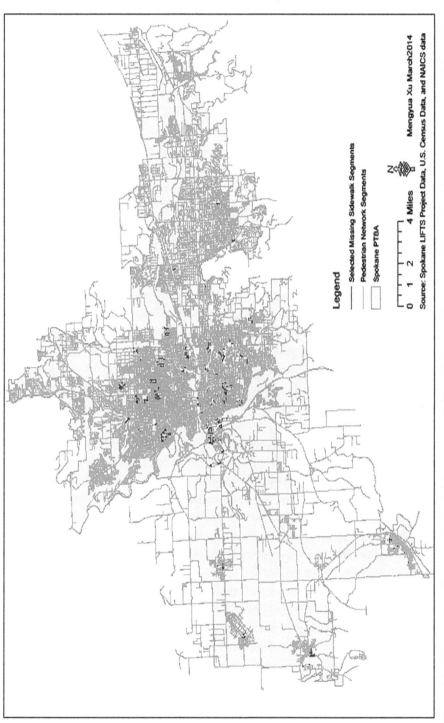

Figure 35: Missing sidewalk segments selected from routes linking all Census block centroids to the closest park entrance within walking distance.

Source: Author.

CHAPTER FOUR   RESULTS

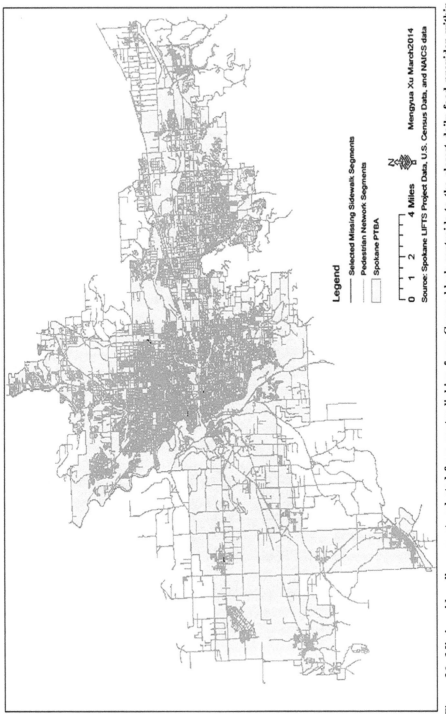

Figure 36: Missing sidewalk segments selected from routes linking focus Census block centroids to the closest daily food provider within walking distance.

Source: Author.

4.3 Intersected Sidewalk from the Routes

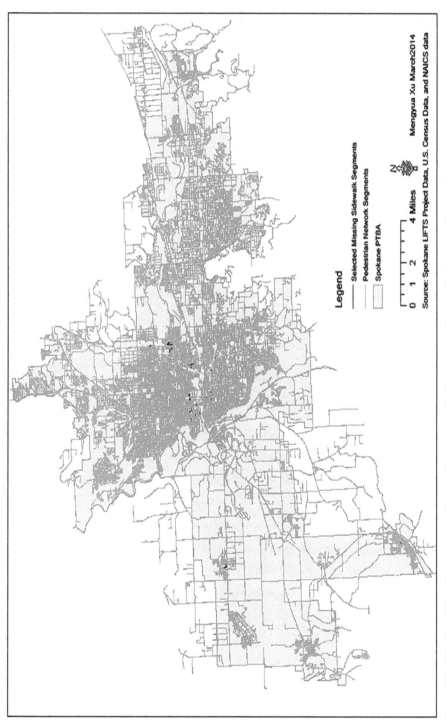

Figure 37: Missing sidewalk segments selected from routes linking focus Census block centroids to the closest gathering place within walking distance.

Source: Author.

CHAPTER FOUR   RESULTS

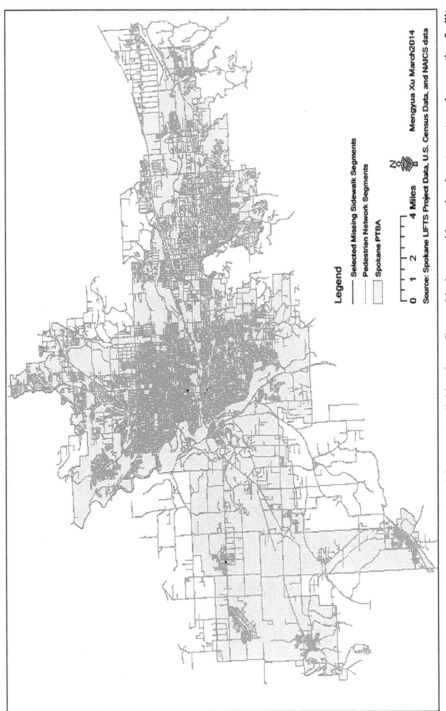

Figure 38: Missing sidewalk segments selected from routes linking focus Census block centroids to the closest sports and recreation facility within walking distance.

Source: Author.

4.3 Intersected Sidewalk from the Routes

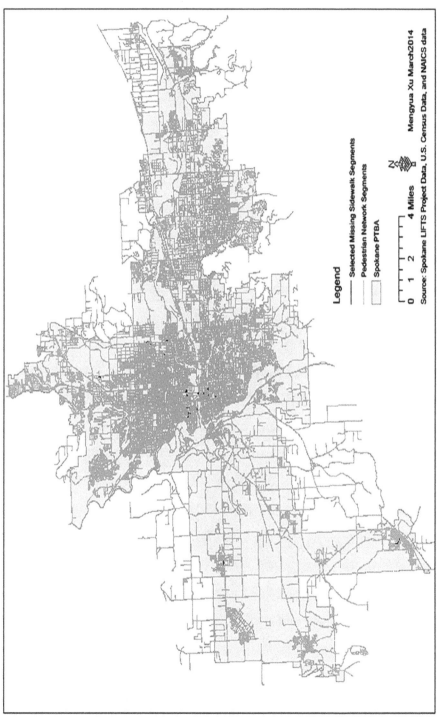

Figure 39: Missing sidewalk segments selected from routes linking focus Census block centroids to the closest healthcare facility within walking distance.

Source: Author.

CHAPTER FOUR   RESULTS

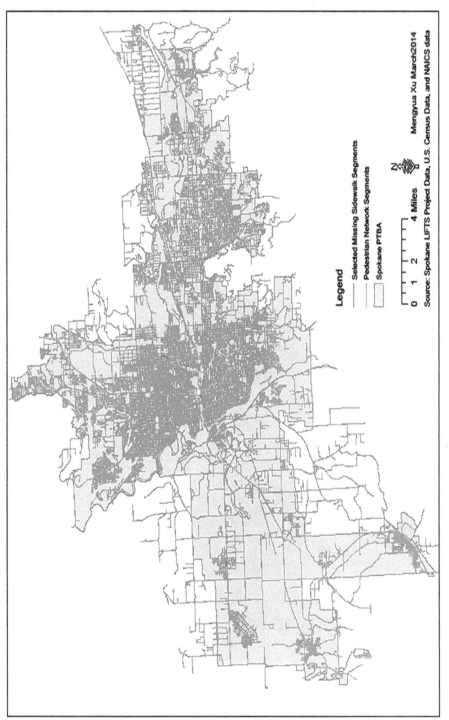

Figure 40: Missing sidewalk segments selected from routes linking focus Census block centroids to the closest park entrance within walking distance.

Source: Author.

4.3 Intersected Sidewalk from the Routes

Based on the intersection of the missing sidewalk segments layer and routes linking all Census block centroids to the daily food provider within walking distance, 256 missing sidewalk segments were selected. The range of the length of the missing sidewalk segments is 0.90 to 618.51 feet with a mean of 147.84 feet. Based on the intersection of the missing sidewalk segments layer and routes linking all Census block centroids to the closest social gathering place within walking distance, 2710 missing sidewalk segments were selected. The range of the length of the missing sidewalk segments is 0.67 to 726.55 feet with a mean of 126.68 feet. Based on the intersection of the missing sidewalk segments layer and routes linking all Census block centroids to the closest sports or recreation facility within walking distance, 341 missing sidewalk segments were selected. The range of the length of the missing sidewalk segments is 2.15 to 769.26 feet with a mean of 157.34 feet. Based on the intersection of the missing sidewalk segments layer and routes linking all Census block centroids to the closest healthcare facility within walking distance, 3202 missing sidewalk segments were selected. The range of the length of the missing sidewalk segments is 0.55 to 938.06 feet with a mean of 126.63 feet. Based on the intersection of the missing sidewalk segments layer and routes linking all Census block centroids to the closest park entrance within walking distance, 256 missing sidewalk segments were selected. The range of the length of the missing sidewalk segments is 0.23 to 1150.32 feet with a mean of 120.47 feet.

Based on the intersection of the missing sidewalk segments layer and routes linking focus Census block centroids to the closest daily food provider within walking distance, 24 missing sidewalk segments were selected. The range of the length of the missing sidewalk segments is 9.54 to 311.77 feet with a mean of 122.69 feet. Based on the intersection of the missing sidewalk segments layer and routes linking focus Census block centroids to the closest social gathering place within walking distance, 108 missing sidewalk segments were selected. The range of the length of the missing sidewalk segments is 9.54 to 342.83 feet with a mean of 137.46 feet. Based on the intersection of the missing sidewalk segments layer and routes linking focus Census block centroids to the closest sport and recreation facility within walking distance, 38 missing sidewalk segments were selected. The range of the length of the missing sidewalk segments is 18.22 to 364.40 feet with a mean of

124.19 feet. Based on the intersection of the missing sidewalk segments layer and routes linking focus Census block centroids to the closest healthcare facility within walking distance, 75 missing sidewalk segments were selected. The range of the length of the missing sidewalk segments is 2.48 to 621.53 feet with a mean of 136.90 feet. Based on the intersection of the missing sidewalk segments layer and routes linking focus Census block centroids to the closest park within walking distance, 77 missing sidewalk segments were selected. The range of the length of the missing sidewalk segments is 5.26 to 311.24 feet with a mean of 106.63 feet.

4.4 Double-Checking Missing Sidewalk Segments on Both Sides of the Roads

Before calculating the overall prioritization, the next step is to calculate the total length of missing sidewalk segments on both sides of the road by each key. If one segment that is selected in the route calculation process has 0 length of total missing sidewalk on the other side of the road based on each KEY value, this segment needs to be eliminated from the overall selection. As discussed in the methodology chapter, the identification process can be conducted based on the data recorded in the road feature class, which can be related to the missing sidewalk segment layer by KEY value.

The result of this analysis is as follows. The highlighted missing sidewalk segments are those eliminated from the overall selection of prioritization. 195 out of a total of 2,541 segments with importance value were identified and eliminated from the final selection. 2,346 missing sidewalk segments are considered as having importance value for improvement.

4.4 Double-Checking Missing Sidewalk Segments on Both Sides of the Roads

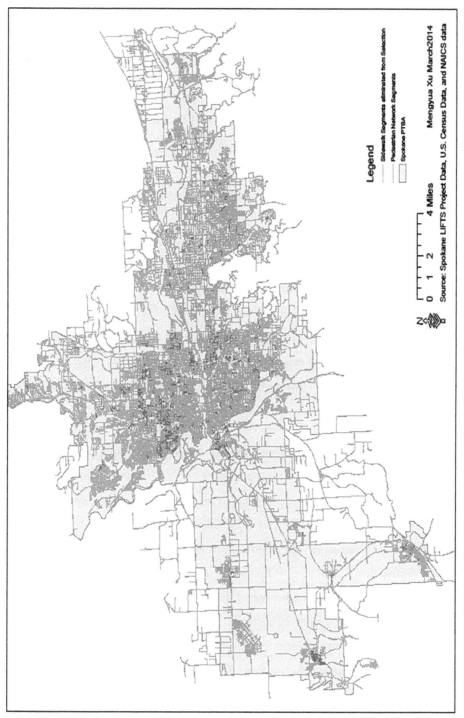

Figure 41: Missing sidewalk segments eliminated from selection.

Source: Author.

4.5 Efficiency and Equity Importance Measure Calculation

The next step is to calculate the Efficiency and Equity Importance Measure for each missing sidewalk segment based on the formulas of the Efficiency Importance Measure and Equity Importance Measure calculation illustrate in Chapter 3 as follows.

$$I_{eff} = (\sum P_{groc} + \sum P_{gat} + \sum P_{spo} + \sum P_{hea} + \sum P_{pa})/L_n$$
$$I_{eq} = (\sum FP_{groc} + \sum FP_{gat} + \sum FP_{spo} + \sum FP_{hea} + \sum FP_{pa})/L_n$$

where

I_{eff} is the Efficiency Importance Measure;

I_{eq} is the Equity Importance Measure;

L_n is the shape length of the missing sidewalk segment;

2346 missing sidewalk segments out of 43981 segments (5.33%) were identified as having an importance value for accessibility improvement. Among these missing sidewalk segments, 75 out of 43981 segments (0.17%) were identified as having an equity importance value larger than 0. These 75 segments and their importance values are presented in Appendix F.

4.6 Prioritization Results of Various Scenarios

Again, the overall prioritization determination for missing sidewalk segments is based on the calculation of combining both the Efficiency Importance Measure and Equity Importance Measure. The equation can be expressed as:

Overall Prioritization Importance = Efficiency Importance + Equity Importance * r

where r = ratio of the emphasis between Efficiency and Equity

The following the final results of sidewalk segment improvement prioritization for various scenarios with different r values.

The last scenario is focusing on the focus Census blocks instead of the whole study area, which means the prioritization is only based on Equity Importance, and the result is shown as Figure 46.

4.6 Prioritization Results of Various Scenarios

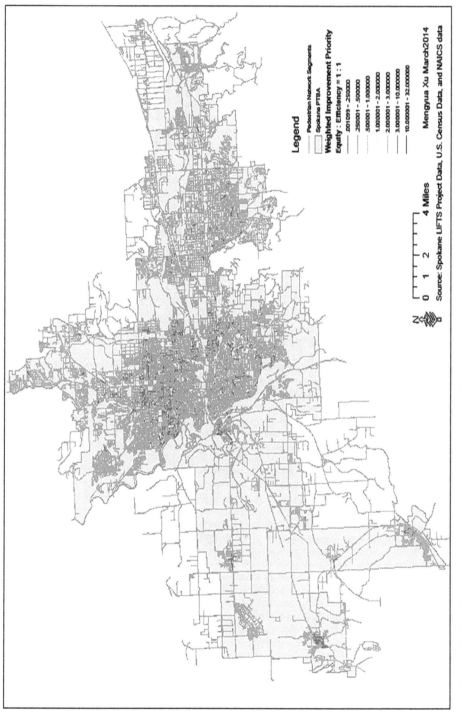

Figure 42: Result of sidewalk segment improvement prioritization (when $r=1$).

Source: Author.

CHAPTER FOUR   RESULTS

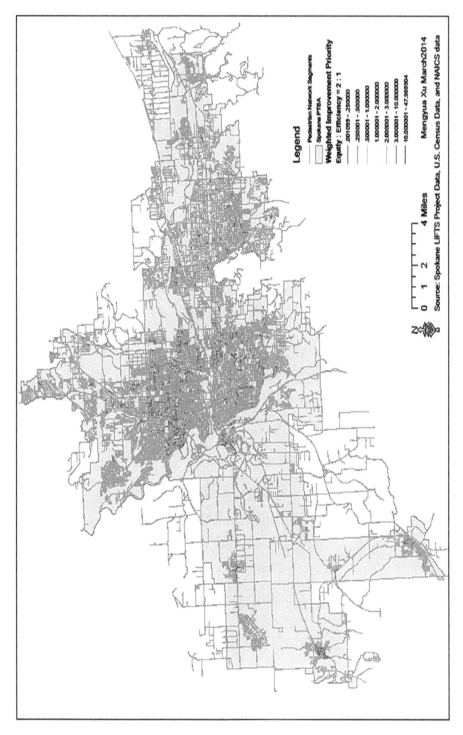

Figure 43: Result of sidewalk segment improvement prioritization (when $r=2$).

Source: Author.

4.6 Prioritization Results of Various Scenarios

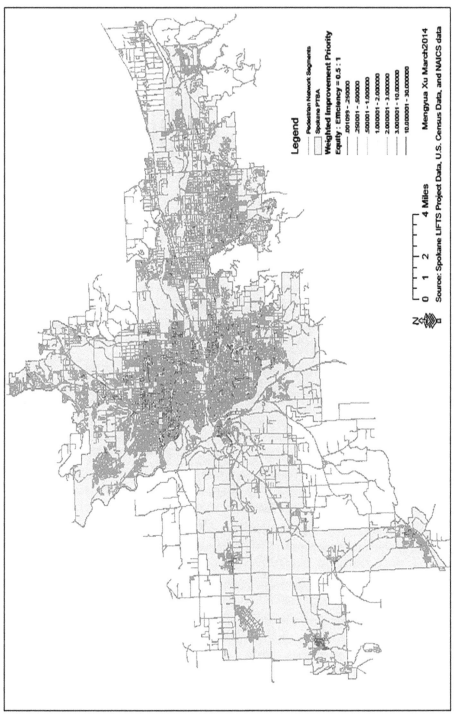

Figure 44: Result of sidewalk segment improvement prioritization (when $r=0.5$).
Source: Author.

CHAPTER FOUR   RESULTS

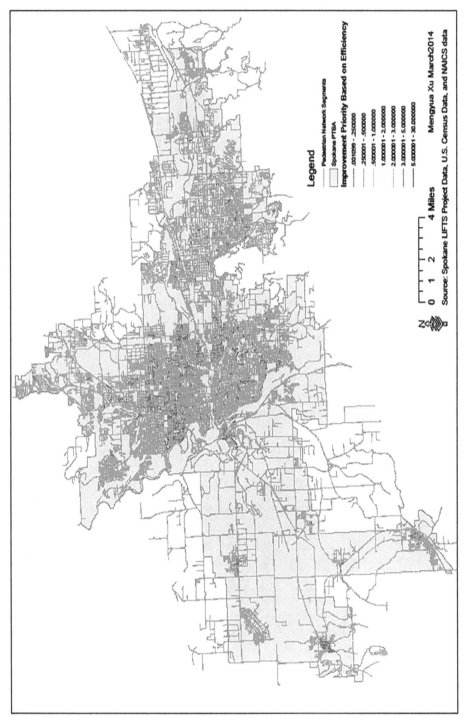

Figure 45: Result of sidewalk segment improvement prioritization (when $r=0$).
Source: Author.

4.6 Prioritization Results of Various Scenarios

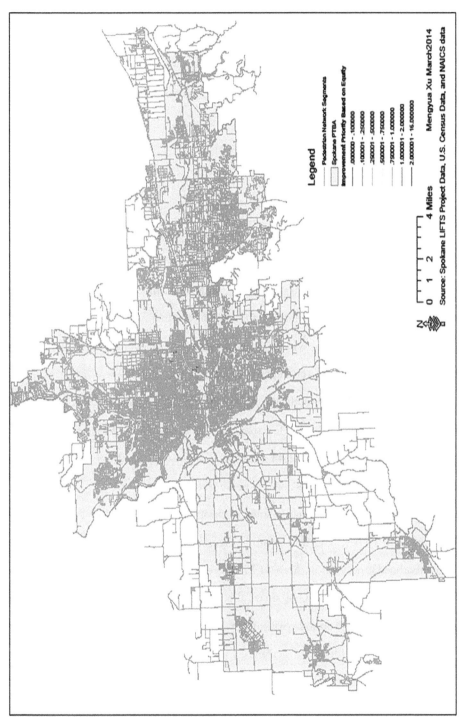

Figure 46: Result of prioritization calculation based on Equity Importance Measure.

Source: Author.

CHAPTER FOUR   RESULTS

The Overall Importance Values for these segments are calculated based on various r values (1, 2, 0.5, and 0). Appendix G, H, I, J presents the selection results of the highest priority segments when r =1, 2, 0.5, and 0, respectively.

When r = 1, the mean value of Overall Importance for these segment is 0.8094 and the standard deviation is 2.0479, with the highest score of 31.9926. When r = 2, the mean value of Overall Importance for these segment is 0.8299 and the standard deviation is 2.1810, with the highest score of 47.9889. When r = 0.5, the mean value of Overall Importance for these segment is 0.7992 and the standard deviation is 2.0000, with the highest score of 29.7469. When r = 0, the mean value of Overall Importance for these segment is 0.7889 and the standard deviation is 1.9668, with the highest score of 29.7469.

From these tables, it can be noticed that the few differences can be found in the results as shown in the Appendix. When r = 1, a total of 42 segments, with a total length of 651.67 feet, are identifies as the highest priority ones, which serve 7597 people in total. These three numbers stay exactly the same when r value is set as 0.5 or 0 (though there are minor differences in details for some of the segments). When r = 2, slight differences can be observed. A total of 44 segments, with a total length of 671.18 feet, are identifies as the highest priority ones, which serve 7663 people in total.

Figure 47 - 54 present the final results of prioritization calculation for two sample areas within Spokane PTBA. The first group of figures shows the results of the area near Hillyard neighborhood (referred as Area #1), where some focus Census blocks have been identified. The second group of figures shows the result of prioritization calculation in the area around the corner of W Nebraska Ave and N Adams St. (referred as Area #2), where no focus Census blocks are identified nearby.

4.6 Prioritization Results of Various Scenarios

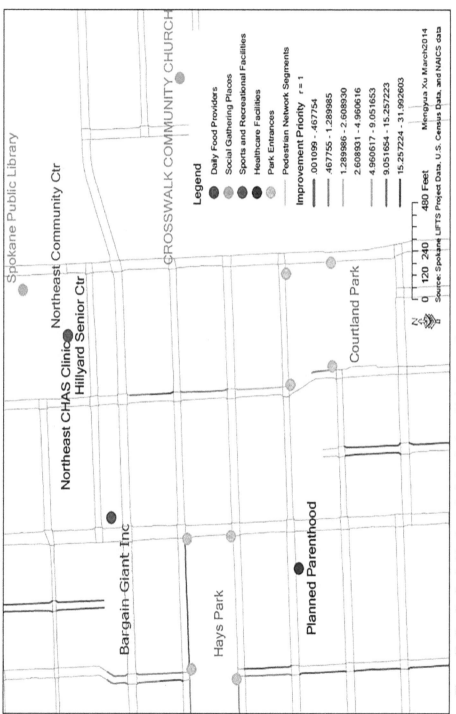

Figure 47: Result of prioritization calculation in Area #1 (when $r=1$).

Source: Author.

CHAPTER FOUR   RESULTS

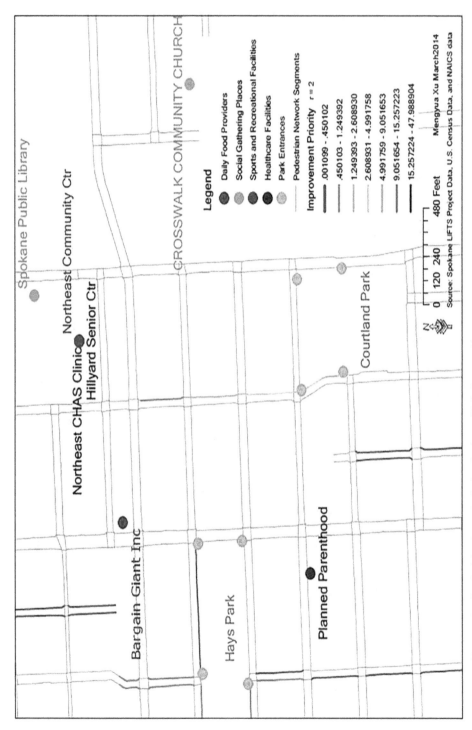

Figure 48: Result of prioritization calculation in Area #1 (when *r*=2).

Source: Author.

4.6 Prioritization Results of Various Scenarios

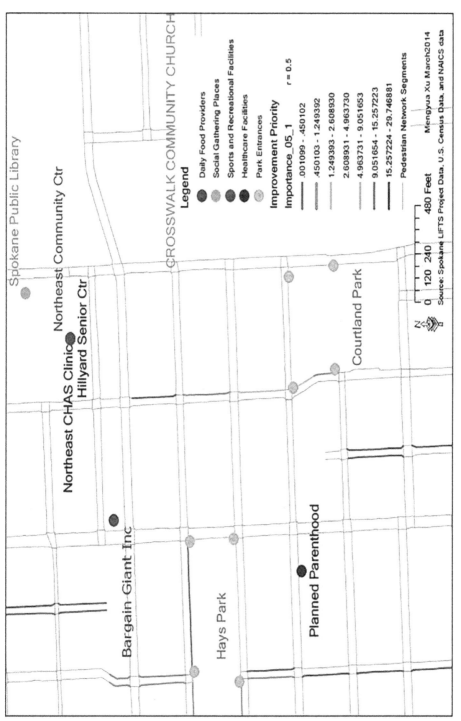

Figure 49: Result of prioritization calculation in Area #1 (when $r=0.5$).
Source: Author.

CHAPTER FOUR RESULTS

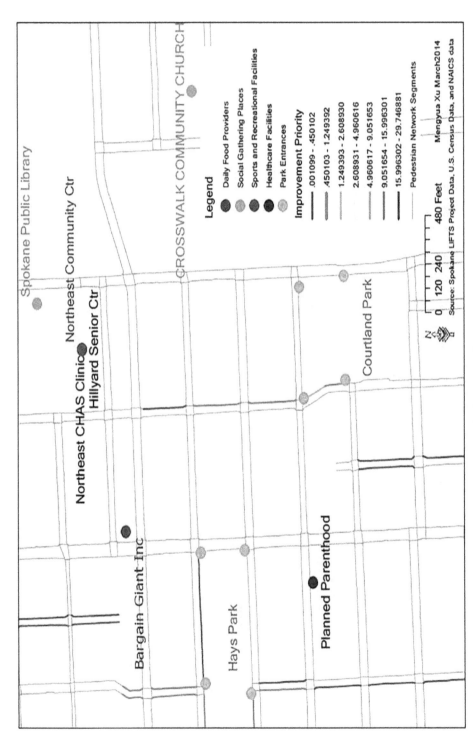

Figure 50: Result of prioritization calculation in Area #1 (when *r* = 0).
Source: Author.

4.6 Prioritization Results of Various Scenarios

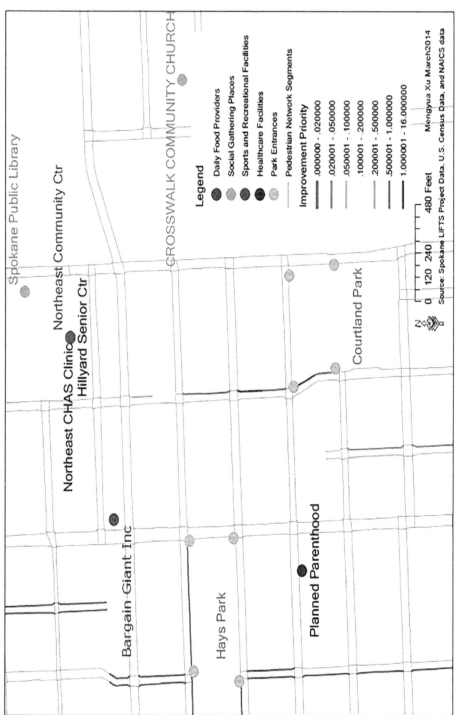

Figure 51: Result of prioritization calculation in Area #1 based on Equity Importance Measure.
Source: Author.

CHAPTER FOUR   RESULTS

Figure 52: Result of prioritization calculation in Area #2 (when $r=1$).
Source: Author.

4.6 Prioritization Results of Various Scenarios

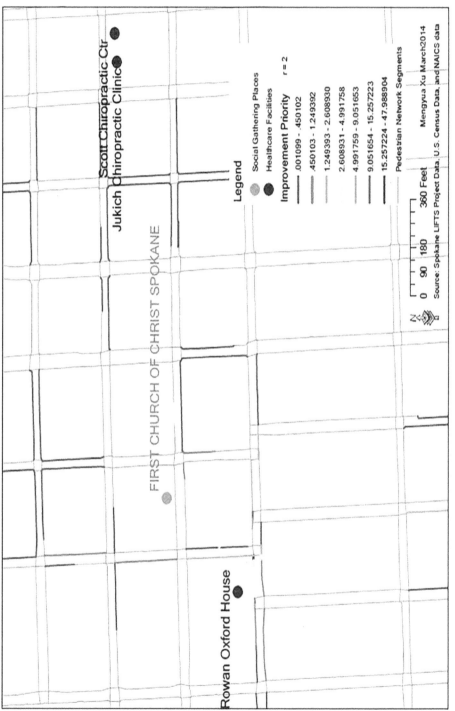

Figure 53: Result of prioritization calculation in Area #2 (when $r=2$).

Source: Author.

CHAPTER FOUR   RESULTS

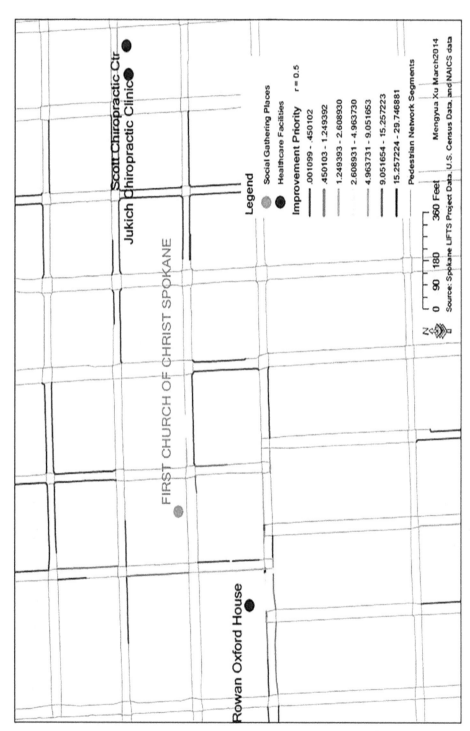

Figure 54: Result of prioritization calculation in Area #2 (when *r*=0.5).
Source: Author.

4.6 Prioritization Results of Various Scenarios

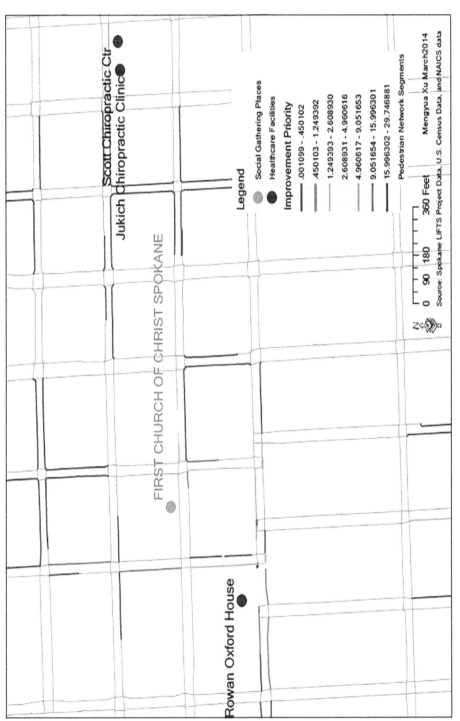

Figure 55: Result of prioritization calculation in Area #2 (when $r=0$).

Source: Author.

CHAPTER FOUR RESULTS

These maps show different prioritization scenarios based on different settings of r values. For the Area #1, different r values have significantly affected results of the prioritization. When r value is set relatively high (2 instead of 1 or 0.5), the overall importance values of missing sidewalk segments near Courtland Park have drastically increased and these segment are considered as high priorities. This change is because this area is close to several focus Census blocks with large number of population and these sidewalk segments locate on the routes connecting these Census blocks to nearby Life-Needs Service Facilities. On the other hand, for the Area #2, different r values have not significantly affected results of the prioritization. For example, the missing sidewalk segments near Rowan Oxford House are considered as high priorities in all scenarios with different r values. This situation is because there is no focus Census block in this area.

CHAPTER FIVE
DISCUSSION AND CONCLUSIONS

5.1 Discussion

Walking behavior and walking environments have drawn increasing research interest recently. Walking is considered as a significant mode of accessing essential urban services for transportation-disadvantaged groups (Litman, 2011). Under these circumstances, pedestrian accessibility analysis plays an increasingly important part in the urban and transport planning policy making process, and measuring pedestrian accessibility has received great interest in recent years (Lotfi & Koohsari, 2009). The main objective of research focusing on measuring the access to urban service facilities is to recommend a more efficient or equitable system (Talen, 2011). In addition, resources can always be limited for infrastructure improvement. As a result, it is crucial for local planning departments to evaluate the effects of potential interventions and decide on the prioritization of the improvement in order to allocate investments in a way that makes every effort optimize the scarce resources based on given design goals and criteria.

Evaluating the impact of complex factors related to accessibility within the pedestrian environment requires robust measurements and analysis techniques. In recent studies, GIS technology has been widely used to perform the measurements and to provide guidance for prioritizing future interventions (Rosero-Bixby, 2004). Currently, the computer-aided design models are at the digital threshold of shifting from representational tools to the next stage of simulation tools which are more actively involved in the design thinking process. In parallel with this concept, Dangermond (2009) from ESRI proposed the concept of GeoDesign, which is a set of design and planning methods that facilitates "the creation of design proposals with impact simulations informed by geographic contexts." Talen (2011, p. 459) found that GIS platform possesses the capability to visualize "the ability to reach urban places" and "the quantity and quality of places that can be reached." These two perspectives comprise the central concept of geovisualization of spatial equity. The basic

CHAPTER FIVE DISCUSSION AND CONCLUSIONS

methodological approach of geovisualization of spatial equity is to map the spatial equity of access and help to understand the potential change of levels of access, which can play an important part as a decision support tool for local government planning (Talen, 2011).

This research proposes a framework of a GIS-based accessibility analysis and decision support tool based upon a comprehensive review of previous studies on accessibility measurements and equity-efficiency trade-off from a theoretical perspective. Drawing concepts from both GeoDesign and Geovisualization, the proposed methodology is novel and fills a gap in previous urban and transport planning literature by drawing from perspectives from multiple disciplines and providing an accessibility assessment model with prioritization tool based on equity and efficiency criteria. This prioritization method is new in the research of this field. This work presents the application of GIS techniques to perform a sophisticated methodology of measuring and mapping pedestrian accessibility in the Spokane PTBA. In order to synthesize the indicators of various aspects in urban service access, the proposed model integrates multiple geospatial data, including the spatial distribution of urban services, socio-demographic data, and transport infrastructure. This research successfully offers a generic method to analyze the level of access to multiple services among different social groups within a certain transport network. The mapping of access routes to these services allows planners and policy makers to locate the Census blocks with no pedestrian access to Life-Needs Service Facilities and thus identify inequities in the level of accessibility across the study area.

In addition, the model provides a mechanism to act as a design decision support tool that enables planners to identify the priority of potential infrastructure improvement based on principles of efficiency and equity. The mapping of the importance of missing segments of a pedestrian network produces recommendations for infrastructure improvement for the Spokane PTBA by identifying the desired interventions with highest priority. Consideration can be placed based on the size of the population that would benefit from the improvement, as well as on the target neighborhoods that require additional planning attention. This prioritization procedure incorporates customized evaluation criteria that could respond to policy makers' various planning goals.

As applied to Spokane PTBA, this model has successfully identified a total of 2346 missing sidewalk segments with Importance value for prioritization. The Overall Importance Values for these segments are calculated based on various r values (1, 2, 0.5, and 0) as shown in Appendix. As mentioned, few differences can be identified with changes of the r values in the selection and prioritization result of highest priority segments. The reason of this situation is that only a small portion of the segments locate on the routes connecting with focus Census blocks and thus have Equity Importance values. As a result, the change of weight of Equity Importance does not affect the final Overall Importance to a great extent. In this situation, I would suggest a high r value (2 or more than 2) in the prioritization process for the study area to put further emphasis on the mobility-disadvantaged neighborhood.

There are some important methodological issues that are worth discussing in the development of the model. First, the present research demonstrates an example of how to develop an accessibility evaluation model through theoretical review of a variety of measurement approaches. As discussed previously, it is proposed that accessibility equity is a three-fold concept. So, the research model for accessibility equity analysis needs to include all three of the following components: social-demographic dimension, land-use/spatial opportunities allocation, and travel option/transportation infrastructure, which connects the previous two. The whole spectrum of accessibility measures was reviewed, and then a proper model structure was chosen for the research topic.

Second, the present model provides a further step in GIS-based quantitative modeling of urban environment simulation and analysis. Many recent researchers have shown interest in exploring the association of urban environments with respect to people's behavior or health. It is suggested that GIS technology has improved the accuracy of measuring urban environments due to its powerful functions (Talen, 2011). However, GIS technology is still new to many researchers, especially those from the healthcare field. The present model provides a robust example of integrating vital factors of urban environment as well as social-demographic data for a comprehensive analysis.

Third, the data collection process of this study illustrated that the geospatial data needed for pedestrian environment analysis could be difficult to derive. The quality of the geospatial data in the analysis model is vital to conduct a reliable accessibility analysis. Using the Network Analyst extension of ESRI software ArcGIS, the proposed model established a network analysis approach that required a detailed pedestrian network dataset. However, there was no available information for sidewalk segments, pedestrian crossings, or other infrastructures related to the pedestrian environment in the study area at the very beginning. The GIS-based pedestrian network datasets used in this study were built with intensive data collection efforts, including rigorous fieldwork and data inputs. On the other hand, although easily available, information for service facilities might not be organized in the desired way and thus needed further processing. For example, raw data on food provider locations were obtained from two NAICS codes (445110 and 452910), which are supermarkets and other grocery stores. However, the latter includes both whole food stores as well as convenience stores that do not provide healthy food (Michimi & Wembley, 2010). As a result, the spatial locations representing convenience stores need to be eliminated from the selection before the accessibility analysis. Similar edits were also conducted for other categories of Life-Needs Service Facilities in advance of accessibility analysis.

Fourth, this study has shown the feasibility of applying GIS technology for decision making in transport planning policy. It represents the trend that GIS has been evolving from a representative tool of describing "what it is" to a concept-stimulation and evaluation tool by predicting "what could be." In other words, previous studies acted as representation tools focusing on a limited function of representing the current conditions, while this research attempts to push the computer-aided design tools in the GIS platform to the next level of a true design decision-supporting tool to help design and evaluate different future scenarios and to find the optimal one by a series of individualized quantitative measures.

To conclude, this study has shown how to establish a theoretical framework for a pedestrian environment model with accessibility measures based on comprehensive literature review. This model organized and operated a complete geospatial information system that performed measurements on the urban service distribution within the transport network. The GIS platform developed in the study can easily juxtapose the supply layer (services locations) with the demand layer (socio-demographic distribution) in the context of a transport network (the Spokane pedestrian network in this study) in order to measure the access within the study area. The data availability and analysis tools make it possible to conduct queries on the impact of potential design interventions and consequently help prioritize improvements to make the distribution of urban services more equitable or efficient.

5.2 Limitations

There are two major limitations in the methodology, and both of them are largely due to data availability. First, the target group was determined by vehicle ownership rate. This data was derived and calculated at the Census block group level because the amount of vehicles at the Census block level or household level (which is even more desirable) was not available in the study area. A dataset on a smaller scale could lead to more accurate analysis results. Second, Life-Needs Service Facilities under one category were all considered identical, without considering the quality of the facilities, or consequently, their potential differences in attractiveness to surrounding citizens. In other words, in this model, every service facility in one category was considered equally competent in fulfilling the needs of people, and thus the infrastructure related to these facilities was considered equally important in prioritization calculation. This assumption could be doubtful, because some facility destinations might be more desirable and frequently visited by people compared with other destinations. The reason for neglecting the differences was the lack of a proper dataset that reflected the desirability of the destinations. Some previous research proposed using the total numbers of customs as an additional criterion. However, this dataset was not readily available for all destinations in the study area. In addition, it is doubtful initially to use total numbers of customs to represent the desirability since it is not necessarily true that more visits necessarily indicate higher importance.

5.3　Future Study

Many issues still need to be researched to provide a more advanced accessibility measurement and design supporting tool to achieve a more equitable or efficient system in the context of transport and urban planning. First, the process of identifying target groups can be extended through other potentially important variables such as income, age, and ethnicity. This research defines the target neighborhoods with car ownership rate, but there are still many other target groups that can draw attention from planning policy makers. More accessibility analysis that integrates the spatial distribution datasets of urban services and various target groups may be able to provide valuable information for proposing future planning scenarios.

Second, the choice and organization of Life-Needs Service Facility categories can also benefit from further improvement based on empirical studies. It was mentioned in the previous chapter that there is no consensus in the literature on deciding the categories of critical urban service destinations. The present research proposes a generic classification that covers a broad spectrum of services. Future studies focusing on certain urban services can develop a more detailed category to fulfill more specific research needs. Organization can be drawn from previous literature of a specific field.

Third, future work can improve the mechanism of the equity-efficiency trade-off. There is a lack of support on how to develop the specific algorithm of the equity-efficiency trade-off from the literature in the field of urban planning. In the present work, the formula of the equity-efficiency trade-off currently used is a linear combination of the two criteria. This formula offers a starting point for a quantitative calculation of the trade-off, but it certainly can be improved from a mathematical standpoint.

BIBLIOGRAPHY

Abley, S. (2005) Walkability scoping paper. Retrieved on March, 4, 2011.

Achuthan, K., Titheridge, H., & Mackett, R. (2010) Mapping accessibility differences for the whole journey and for socially excluded groups of people. *Journal of Maps*, 6 (1), 220-229.

Algert, S. J., Agrawal, A., & Lewis, D. S. (2006) Disparities in access to fresh produce in low-income neighborhoods in Los Angeles. *American Journal of Preventive Medicine*, 30(5), 365-370.

American Community Survey Data. (2010) Retrieved May, 2013.

Apparicio, P., Abdelmajid, M., Riva, M., & Shearmur, R. (2008) Comparing alternative approaches to measuring the geographical accessibility of urban health services: Distance types and aggregation-error issues. *International Journal of Health Geographics*, 7(1), 476-489.

Arentze, T. A., Borgers, A. W. J., & Timmermans, H. J. P. (1994) Geographical Information Systems and the measurement of accessibility in the context of multipurpose travel: A new approach. *Geographical System*, 1, 87-102.

Araya, R., Dunstan, F., Playle, R., Thomas, H., Palmer, S., & Lewis, G. (2006) Perceptions of social capital and the built environment and mental health. *Social Science & Medicine*, 62(12), 3072-3083.

Bader, M. D. M., Purciel, M., Yousefzadeh, P., & Neckerman, K. M. (2010) Disparities in neighborhood food environments: Implications of measurement strategies. *Economic Geography*, 86(4), 409-430.

Beaulac, J., Kristjansson, E., & Cummins, S. (2009) A Systematic Review of Food Deserts, 1966-2007. *Preventing Chronic Disease*, 6(3), 105-114.

Bergman, S. E. (1998) Swedish models of health care reform: a review and assessment. *The International Journal of Health Planning and Management*, 13(2), 91-106.

Birch, E. L. (1980) Radburn and the American Planning Movement The Persistence of an Idea. *Journal of the American Planning Association*, 46(4), 424-439.

Blanchard, W. (1986) Evaluating Social Equity: What does fairness mean and can we measure it? *Policy Studies Journal*, 15(1), 29-54.

Bleichrodta, H., Doctorb, J., & Stolk, E. (2004) A nonparametric elicitation of the equity-efficiency trade-off in cost-utility analysis. *Journal of Health Economics*, 24(4), 655-678.

Bolton, P., & Dewatripont, M. (2005) *Contract theory*. MIT Press.

Bourguignon, F., Ferreira, F. H., & Walton, M. (2007) Equity, efficiency and inequality traps: A research agenda. *Journal of Economic Inequality*, 5(2), 235-256.

Brooks, K. (2013) *Spokane Pedestrian Network*. Washington State University GIS and Simulation Laboratory

Brownson, R. C., Hoehner, C. M., Day, K., Forsyth, A., & Sallis, J. F. (2009) Measuring the built environment for physical activity. *American Journal of Preventive Medicine*, 36(4), 99-132.

Bums, L.D. (1979) Transportation: Temporal and Spatial Components of Accessibility. *Lexington Books*, Lexington, MA.

Casey, S. (2005) *Establishing standards for social infrastructure*. UQ Boilerhouse, Community Engagement Centre.

Cervero, R., & Duncan, M. (2003) Walking, bicycling, and urban landscapes: evidence from the San Francisco Bay Area. *American Journal of Public Health*, 93(9), 1478-1483.

Chatman, D. G. (2009) Residential choice, the built environment, and nonwork travel: evidence using new data and methods. *Environment and Planning. A*, 41(5), 1072.

Chin, G., Vanniel, K., Gilescorti, B., & Knuiman, M. (2008) Accessibility and connectivity in physical activity studies: The impact of missing pedestrian data. *American Journal of Preventive Medicine*, 46(1), 41-45.

Church, R. L., & Marston, J. R. (2003) Measuring Accessibility for People with a Disability. *Geographical Analysis*, 35(1), 83-96.

Cohen, G. A. (1989) On the currency of egalitarian justice. *Ethics*, 99, 906-944.

Cohen, G. A. (1990) Equality of what? On welfare, goods and capabilities. Université catholique de Louvain, Institut de Recherches Economiques et Sociales (IRES).

Comber, A., Brunsdon, C., & Green, E. (2008) Using a GIS-based network analysis to determine urban greenspace accessibility for different ethnic and religious groups. *Landscape and Urban Planning*, 86(1), 103-114.

Connor, S., & Brink, S. (1999) Understanding the early years. *Community Impacts on Child Development*.

Crompton, J. L., & Wicks, B. E. (1988) Implementing a preferred equity model for the delivery of leisure services in the US context. *Leisure Studies*, 7(3), 287-304.

Cummins, S. (2007) Neighbourhood food environment and diet: time for improved conceptual models? *Journal of Preventive Medicine*, 44(3), 196-197.

Cummins, S., & Macintyre, S. (2002) "Food deserts"—evidence and assumption in health policy making. *British Medical Journal*, 325(7361), 436-438.

Dangermond, J. (2009) GIS: Designing our future. *ArcNews*, 31(2), 1-9.

De Vries S., Verheij, R.A., Groenewegen, P.P., & Spreeuwenberg, P. (2003) Natural environments — healthy environments? An exploratory analysis of the relationship between greenspace and health. *Environment and Planning*, 35(10), 1717-1731.

Dietz, S. & Atkinson, G. (2010) The equity-efficiency trade-off in environmental policy: Evidence from stated preferences. *Land Economics*, 86 (3), 423-443.

Dworkin, R. (1981a) What is equality? Part I: Equality of welfare. *Philosophy and Public Affairs*, 10, 185-246.

Dworkin, R. (1981b) What is equality? Part 2: Equality of resources. *Philosophy and Public Affairs*, 10, 283-345.

Duany, A., Plater-Zyberk, E., & Speck, J. (2001) *Suburban nation: The rise of sprawl and the decline of the American dream*. Macmillan.

Egan, J. (2004) The Egan review: skills for sustainable communities.

Erikson, R., & Goldthorpe, J. H. (1992) The CASMIN project and the American dream. *European Sociological Review*, 8(3), 283-305.

Ewing, R., Handy, S., Brownson, R. C., Clemente, O., & Winston, E. (2006) Identifying and measuring urban design qualities related to walkability. *Journal of Physical Activity and Health*, 3(1), 223-240.

Eyler, A. A., Brownson, R. C., Bacak, S. J., & Housemann, R. A. (2003) The epidemiology of walking for physical activity in the United States. *Medicine and Science in Sports and Exercise*, 35(9), 1529-1536.

Ford, P. B., & Dzewaltowski, D. A. (2008) Disparities in obesity prevalence due to variation in the retail food environment: three testable hypotheses. *Nutrition Reviews*, 66(4), 216-28.

Gebel, K., Bauman, A. E., and Petticrew, M. (2007) The physical environment and physical activity: a critical appraisal of review articles. *American Journal of Preventive Medicine*, 32(5), 361-369.

Geurs, K., Krizek, K., & Reggiani, A. (2012) Accessibility Analysis and Transport Planning — Challenges for Europe and North America. Edward Elgar Publishing Limited.

Geurs, K. T., & Ritsema van Eck, J. R. (2001) Accessibility measures: review and applications. Evaluation of accessibility impacts of land-use transport scenarios, and related social and economic impacts. RIVM —National Institute of Public Health and the Environment, Bilthoven.

Giang, T., Karpyn, A., Laurison, H. B., Hillier, A., & Perry, R.D. (2008) Closing the grocery gap in underserved communities: The creation of the Pennsylvania Fresh Food Financing Initiative. *Journal of Public Health Management and Practice*, 14 (3), 272-279.

Grafova, I. B. (2008) Overweight children: Assessing the contribution of the built environment. *American Journal of Preventive Medicine*, 47, 304-308.

Groat, L. and Wang, D, (2002) *Architectural Research Methods*. Wiley, New York.

Groenewegen, P. P., van den Berg, A. E., de Vries, S., & Verheij, R. A. (2006) Vitamin G: effects of green space on health, well-being, and social safety.

Gutierrez, J., Condeco-Melhorado, A., & Martin, J.C. (2010) Using accessibility indicators and GIS to assess spatial spillovers of transport infrastructure investment. *Journal of Transport Geography*, 18(1), 141-152.

Handy, S. L., Niemeier, D. A. (1997) Measuring accessibility: an exploration of issues and alternatives. *Environment and Planning*, 29(7), 1175-1194.

Handy, S.L., & Clifton, K.J. (2001) Evaluating neighborhood accessibility: Possibilities and practicalities. *Journal of Transportation and Statistics*, 4(2), 67-78.

Hansen, W. G. (1959). How accessibility shapes land use. *Journal of American Institute of Planners*, 25(2), 73-76.

Hanson, S., & Pratt, G. (1988) Reconceptualizing the links between home and work in urban geography. *Economic Geography*, 299-321.

Haskell, W. L., Lee, I. M., Pate, R. R., Powell, K. E., Blair, S. N., Franklin, B. A., ... & Bauman, A. (2007) Physical activity and public health: updated recommendation for adults from the American College of Sports Medicine and the American Heart Association. *Medicine and Science in Sports and Exercise*, 39(8), 1423-1434.

Hay, A. M. (1995) Concepts of equity, fairness and justice in geographical studies. *British Geographers*, 20(4), 500-508.

Herzog, T.R., & Strevey, S.J. (2008) Contact with nature, sense of humor, and psychological well-being. *Environment and Behavior*, 40(6), 747-776.

Higgs, G. (2005) A literature review of the use of GIS-base measures of access to health care services. *Health Services & Outcomes Research Methodology*, 5(2), 119-136.

Hoehner, C. M., Brennan Ramirez, L. K., Elliott, M. B., Handy, S. L., & Brownson, R. C. (2005) Perceived and objective environmental measures and physical activity among urban adults. *American Journal of Preventive Medicine*, 28(2), 105-116.

Humpel, N., Owen, N., & Leslie, E. (2002) Environmental factors associated with adults' participation in physical activity: a review. *American Journal of Preventive Medicine*, 22(3), 188-199.

Iacono, M., Krizek, K., & El-Geneidy, A. (2010) Measuring non-motorized accessibility: Issues, alternatives, and execution. *Journal of Transport Geography*, 18(1), 133-140.

INFO USA Data. (2012). Retrieved May, 2013.

Jacobson, J. O., Hengartner, N. W., & Louis, T. A. (2004) Inequity measures for evaluations of environmental justice: a case study of close proximity to highways in NYC. *Johns Hopkins University, Dept. of Biostatistics Working Papers*. Working Paper 29.

Kaplan, S. (1995) The restorative benefits of nature: Toward an integrative framework. *Journal of Environmental Psychology*, 15(3), 169-182.

Karner, A., & Niemeier, D. (2013) Civil rights guidance and equity analysis methods for regional transportation plans: a critical review of literature and practice. *Journal of Transport Geography*, 33, 126-134.

Kelly, C. E., Tight, M. R., Hodgson, F. C., & Page, M. W. (2010) A comparison of three methods for assessing the walkability of the pedestrian environment. *Journal of Transport Geography*, 8(1), 1016-1024.

Knapp, K. K., & Hardwick, K. (2000) The availability and distribution of dentists in rural ZIP codes and primary care health professional shortage areas (PC-HPSA) ZIP codes: comparison with primary care providers. *Journal of Public Health Dentistry*, 60(1), 43-48.

Knox, E. G. (1978) Principles of allocation of health care resources. *Journal of Epidemiology and Community Health*, 32(1), 3-9.

Kotnik, T. (2010) Digital architectural design as exploration of computable functions. *International Journal of Architectural Computing*, 8(1), 1-16.

Krumholz, N. (1982) A retrospective view of equity planning Cleveland 1969-1979. *Journal of the American Planning Association*, 48(2), 163-174.

Larsen, K. & Gilliland, J. (2008) Mapping the evolution of 'food deserts' in a Canadian city: Supermarket accessibility in London, Ontario, 1961-2005. *International Journal of Health Geographics*, 7(1), 16.

Lasser, K. E., Himmelstein, D. U., & Woolhandler, S. (2006) Access to care, health status, and health disparities in the United States and Canada: results of a cross-national population-based survey. *American Journal of Public Health*, 96(7), 1300.

le Grand, J. (1990) Equity versus efficiency: The elusive trade-off. *Ethis*, 100(3), 554-568. Lee, C. & Moudon, A. V. (2006). Correlates of walking for transportation or recreation purposes. *Journal of Physical Activity and Health*, 3(1), 77-98.

Leslie, E., Coffee, N., Frank, L., Owen, N., Bauman, A., & Hugo, G. (2007) Walkability of local communities: Using geographic information systems to objectively assess relevant environmental attributes. *Health and Place*, 13(1), 111-122.

Levinson, D. M., & Krizek, K. l. (2008) *Planning for Place and Plexus*. New York: Routledge.

Leyden, K. (2003) Social Capital and the Built Environment: The Importance of Walkable Neighborhoods. *Research and Practice*, 93(9), 1546-1551.

Litman, T. (2002) Evaluating Transportation Equity. *World Transport Policy & Practice*, 8(2), 50-65.

Litman, T. (2003a) Economic value of walkability. *Transportation Research Record*, 10 (1), 3-11.

Litman, T. (2003b) Social inclusion as a transport planning issue in Canada. In *Transport and Social Exclusion G7 Comparison Seminar*. London.

Litman, T. (2008). Evaluating accessibility for transportation planning. *Victoria Transport Policy Institute*, Victoria, Canada.

Litman, T., & Brenman, M. (2012) *A New Social Equity Agenda for Sustainable Transportation*. Victoria Transport Policy Institute.

Livi, A. D. & Clifton, K. J. (2004) Issues and methods in capturing pedestrian behaviors, attitudes and perceptions: Experiences with a community-based walkability survey. 2004 USA Transportation Research Board Meeting.

Lotfi, S., & Koohsari, M. J. (2009) Analyzing accessibility dimension of urban quality of life: Where urban designers face duality between subjective and objective reading of place. *Social Indicators Research*, 94(3), 417-435.

Lopez, R. P. (2007) Neighborhood risk factors for obesity. *Obesity*, 15(8), 2111-2119.

Lucy, W. (1981) Equity and planning for local services. *Journal of the American Planning Association*, 47(4), 447-457.

Luo, W., & Wang, F. (2003) Measures of spatial accessibility to health care in a GIS environment: Synthesis and a case study in the Chicago region. *Environment Planning and Design*, 30(6), 865-884.

Manderscheid, K. (2012) Planning sustainability: Intergenerational and intragenerational justice in spatial planning strategies. *Antipode*, 44(1), 197-216.

Martens, K. (2012) Justice in transport as justice in access: applying Walzer's 'Spheres of Justice' to the transport sector. *Transportation*, 39(6), 1035-1053.

Martens, K., Golub, A., & Robinson, G. (2012) A justice-theoretic approach to the distribution of transportation benefits: Implications for transportation planning practice in the United States. *Transportation Research Part A: Policy and Practice*, 46(4), 684-695.

McMaster, R.B., Leitner, H., Sheppard, E., (1997) GIS-based environmental equity and risk assessment: methodological problems and prospects. *Cartography and Geographic Information Systems*, 24(3), 172-189.

Michimi, A., & Wimberly, M.C. (2010) Associations of supermarket accessibility with obesity and fruit and vegetable consumption in the conterminous United States. *International Journal of Health Geographics*, 9(1), 49-62.

Millington, C., Ward, T.C., Aspinall, P., Rowe, D., Fitzsimons, C., Nelson, N., & Mutrie, N. (2009) Development of the Scottish walkability assessment tool (SWAT). *Health and Place*, 15(2), 474-481.

Mladenka, K.R. (1980) The urban bureaucracy and the Chicago political machine: Who gets what and the limits to political control. *The American Political Science Review*, 74(4), 991-998.

Moore, L. V., Diez Roux, A. V., Nettleton, J. A., & Jacobs, D. R. (2008) Associations of the local food environment with diet Quality—A comparison of assessments based on surveys and geographic information systems: The multi-ethnic study of atherosclerosis. *American Journal of Epidemiology*, 167(8), 917-924.

Morland, K., Diez Roux, A. V., & Wing, S. (2006) Supermarkets, other food stores, and obesity: The atherosclerosis risk in communities study. *American Journal of Preventive Medicine*, 30(4), 333-339.

Morland, K. B., & Evenson, K. R. (2009) Obesity prevalence and the local food environment. *Health & Place* 15(2), 491-495.

Moudon, A.V., Lee, C., Cheadle, A.D., Garvin, C., Johnson, D., Schmid, T.L., Weathers, R. D., & Lin, L. (2006) Operational definitions of walkable neighborhood: Theoretical and empirical insights. *Journal of Physical Activity and Health*, 3(1), 99-117.

Neutens, T., Schwanen, T., Witlox, F., & De Maeyer, P. (2010) Equity of urban service delivery: a comparison of different accessibility measures. *Environment and planning. A*, 42(7), 1613-1635.

Ngui, A. N., & Philippe Apparicio, P. (2011) Optimizing the two-step floating catchment area method for measuring spatial accessibility to medical clinics in Montreal. *BMC Health Services Research*, 11(1), 166,

Ohnmacht, T., Maksim, H., & Bergman, M. M. (Eds.) (2009) *Mobilities and inequality*. Ashgate Publishing, Ltd..

Paranagamage, P., Austin, S., Price, A., & Khandokar, F. (2010) Social capital in action in urban environments: an intersection of theory, research and practice literature. *Journal of Urbanism: International Research on Placemaking and Urban Sustainability*, 3(3), 231-252.

Pikora, T. J., Bull, F.C., Jamrozik, K., Knuiman, M., Giles-Corti, B., & Donovan, R. J. (2002) Developing a reliable audit instrument to measure the physical environment for physical activity. American *Journal of Preventive Medicine*, 23(3), 187-194.

Ploeg, M. V., Breneman, V., Farrigan, T., Hamrick, K., Hopkins, D., Kaufman, P., ... & Tuckermanty, E. (2009) Access to affordable and nutritious food: measuring and understanding food deserts and their consequences. Report to Congress. In *Access to affordable and nutritious food: measuring and understanding food deserts and their consequences. Report to Congress*. USDA Economic Research Service.

Pothukuchi, K. (2005) Attracting Supermarkets to Inner-City Neighborhoods: Economic Development Outside the Box. *Economic Development Quarterly*, 19(3), 232-244.

Powell, L. M., Auld, M. C., Chaloupka, F. J., O'Malley, P. M., & Johnston, L. D. (2007) Associations between access to food stores and adolescent body mass index. *American Journal of Preventive Medicine*, 33(4), 301-307.

Rawls, J. (1971) *A Theory of Justice*. Harvard University Press, Cambridge, MA.

Rose, D., & Richards, R. (2004) Food store access and household fruit and vegetable use among participants in the US Food Stamp Program. *Public Health Nutrition*, 7(1), 1081-1088.

Rosero-Bixby, L. (2004) Spatial access to health care in Costa Rica and its equity: A GIS-based study. *Social Science and Medicine*, 58, 1271-1284.

Saelens, B. E. & L. Handy, S. L. (2008) Built environment correlates of walking: a review. *Medicine & Science in Sports & Exercise*, 40(7), 550-566.

Salze, P., Banos, A., Oppert, J. M., Charreire, H., Casey, R., Simon, C., ... & Weber, C. (2011) Estimating spatial accessibility to facilities on the regional scale: an extended commuting-based interaction potential model. *International Journal of*

Health Geographics, 10(1), 2-17.

Grethe B. Peterson (Ed.). (2011) *The Tanner lectures on human values* (Vol. 9). Cambridge University Press.

Shaw, H. J. (2006) Food deserts: towards the development of a classification. *Geografiska Annaler: Series B, Human Geography*, 88(2), 231-47.

Smith, C. (1994) The new corporate philanthropy. *Harvard Business Review*, 72(3), 105-116.

Smoyer-Tomic, K.E., Hewko, J.N., & Hodgson, M.J. (2004) Spatial accessibility and equity of playgrounds in Edmonton, Canada. *The Canadian Geographer*, 48(3), 287-302.

Susi, L., & Mascarenhas, A.K. (2002) Using a geographical information system to map the distribution of dentists in Ohio. *Journal of the American Dental Association*, 133(5), 636-642.

Talen, E. (2011) Geovisualization of Spatial Equity. *The SAGE Handbook of GIS and Society*, 458-479.

Talen, E. (2000) The problem with community in planning. *Journal of Planning Literature*, 15(2), 171-183.

Talen, E. (1996) The social equity of urban service distribution: An exploration of park access in Pueblo, Colorado, and Macon, Georgia. *Urban Geography*, 18(6), 521-553.

Ulrich, R. S. (1979) Visual landscapes and psychological well-being. *Landscape Research*, 4(1), 17-23.

U.S. Census Bureau. (2013) Annual estimates of the resident population for incorporated places over 50,000. Available at: http://factfinder2.Census.gov/bkmk/table/1.0/en/PEP/2012/PEPANNRSIP.US12A

Van Bergeijk, E., Bolt, G., & Van Kempen, R. (2008) Social cohesion in deprived neighborhoods in the Netherlands: The effect of the use of neighborhood facilities. In Annual Meeting of the Housing Studies Association, York, UK.

Vandenbulcke, G., Steenberghen, T., & Thomas, I. (2009) Mapping accessibility in Belgium: A tool for land-use and transport planning? *Journal of Transport Geography*, 17(1), 39-53.

Walker, R. E., Keane, C. R., & Burke, J. G. (2010) Disparities and access to healthy food in the United States: a review of food deserts literature. *Health & Place*, 16(5), 876-884.

Weber, J. & Kwan, M. (2003) Evaluating the effects of geographic contexts on

Individual accessibility: a multilevel approach. *Urban Geography*, 24(8), 647-671.

Weibull, J. W. (1980) On the numerical measurement of accessibility. *Environment Planning*, 12(1), 53-67.

Yigitcanlar, T. (2007) A GIS-based land use and public transport accessibility indexing model. *Australian Planner*, 44(3), 30-37.

Zenk, S. N., Schulz, A. J., Israel, B. A., James, S. A., Bao, S., & Wilson, M. L. (2005) Neighborhood racial composition, neighborhood poverty, and the spatial accessibility of supermarkets in metropolitan Detroit. *American Journal of Public Health*, 95(4), 660-667.

APPENDIX

Appendix A: List of all Daily Food Providers within the Study Area

DesID	COMPANY_NA	SELECTED_1	NAICS_CODE	NAICS_DESC
1100	Bongs Grocery & Deli	Grocers-Retail	44511003	Supermarkets & Other Grocery Stores
2100	Broadway Foods	Grocers-Retail	44511003	Supermarkets & Other Grocery Stores
3100	Family Superette	Grocers-Retail	44511003	Supermarkets & Other Grocery Stores
4100	P M Jacoy's	Grocers-Retail	44511003	Supermarkets & Other Grocery Stores
5100	Rick's Grocery	Grocers-Retail	44511003	Supermarkets & Other Grocery Stores
6100	Rosauers Supermarkets Inc.	Grocers-Retail	44511003	Supermarkets & Other Grocery Stores
7100	Bollywood Video & Groceries	Grocers-Retail	44511003	Supermarkets & Other Grocery Stores
8100	Diane's Foods Inc.	Food Products-Retail	44511002	Supermarkets & Other Grocery Stores
9100	Fred Meyer	Grocers-Retail	44511003	Supermarkets & Other Grocery Stores
10100	Grocery Outlet	Grocers-Retail	44511003	Supermarkets & Other Grocery Stores
11100	Hamilton Mart	Food Markets	44511001	Supermarkets & Other Grocery Stores
12100	International Foods Store	Grocers-Retail	44511003	Supermarkets & Other Grocery Stores
13100	Kraft Foods	Food Products-Retail	44511002	Supermarkets & Other Grocery Stores
14100	Piccolo's Market	Grocers-Retail	44511003	Supermarkets & Other Grocery Stores
15100	Washington Food Policy Action	Food Products-Retail	44511002	Supermarkets & Other Grocery Stores
16100	Albertsons	Grocers-Retail	44511003	Supermarkets & Other Grocery Stores
17100	Rocket Market	Grocers-Retail	44511003	Supermarkets & Other Grocery Stores
18100	Super 1 Foods	Grocers-Retail	44511003	Supermarkets & Other Grocery Stores
19100	Rosauers Supermarkets Inc.	Grocers-Retail	44511003	Supermarkets & Other Grocery Stores
20100	Albertsons	Grocers-Retail	44511003	Supermarkets & Other Grocery Stores
21100	J B's Foods	Grocers-Retail	44511003	Supermarkets & Other Grocery Stores

APPENDIX

Appendix A cont'd

DesID	COMPANY_NA	SELECTED_1	NAICS_CODE	NAICS_DESC
22100	Priced-Rite Foods	Grocers-Retail	44511003	Supermarkets & Other Grocery Stores
23100	Rosauers Supermarkets Inc.	Grocers-Retail	44511003	Supermarkets & Other Grocery Stores
24100	Safeway	Grocers-Retail	44511003	Supermarkets & Other Grocery Stores
25100	Safeway	Grocers-Retail	44511003	Supermarkets & Other Grocery Stores
26100	Sure Save Grocery	Grocers-Retail	44511003	Supermarkets & Other Grocery Stores
27100	Antonio's Pizza	Grocers-Retail	44511003	Supermarkets & Other Grocery Stores
28100	Bargain Giant Inc.	Grocers-Retail	44511003	Supermarkets & Other Grocery Stores
29100	Empire Food	Grocers-Retail	44511003	Supermarkets & Other Grocery Stores
30100	Fresh Abundance	Grocers-Retail	44511003	Supermarkets & Other Grocery Stores
31100	Kiev Market	Grocers-Retail	44511003	Supermarkets & Other Grocery Stores
32100	M & K Grocery	Grocers-Retail	44511003	Supermarkets & Other Grocery Stores
33100	Mike's Grocery	Grocers-Retail	44511003	Supermarkets & Other Grocery Stores
34100	Sure Save Grocery	Grocers-Retail	44511003	Supermarkets & Other Grocery Stores
35100	Wynn Starr Foods Inc-Kentucky	Food Products-Retail	44511002	Supermarkets & Other Grocery Stores
36100	Yoke's Fresh Market	Grocers-Retail	44511003	Supermarkets & Other Grocery Stores
37100	Albertsons	Grocers-Retail	44511003	Supermarkets & Other Grocery Stores
38100	Albertsons	Grocers-Retail	44511003	Supermarkets & Other Grocery Stores
39100	Cash & Carry	Grocers-Retail	44511003	Supermarkets & Other Grocery Stores
40100	De Leon Foods Inc.	Grocers-Retail	44511003	Supermarkets & Other Grocery Stores
41100	G & B Grocery	Grocers-Retail	44511003	Supermarkets & Other Grocery Stores
42100	Grocery Outlet	Grocers-Retail	44511003	Supermarkets & Other Grocery Stores
43100	Hillyard Lockers	Grocers-Retail	44511003	Supermarkets & Other Grocery Stores
44100	Yoke's Foods	Grocers-Retail	44511003	Supermarkets & Other Grocery Stores
45100	Rosauers Supermarkets Inc.	Grocers-Retail	44511003	Supermarkets & Other Grocery Stores
46100	Sunshine Dairy	Grocers-Retail	44511003	Supermarkets & Other Grocery Stores
47100	Sunshine Dairy Inc.	Grocers-Retail	44511003	Supermarkets & Other Grocery Stores
48100	Safeway	Grocers-Retail	44511003	Supermarkets & Other Grocery Stores
49100	Akins Foods Inc.	Grocers-Retail	44511003	Supermarkets & Other Grocery Stores
50100	Rosauers Supermarkets Inc	Grocers-Retail	44511003	Supermarkets & Other Grocery Stores
51100	Tidyman's	Grocers-Retail	44511003	Supermarkets & Other Grocery Stores
52100	Albertsons	Grocers-Retail	44511003	Supermarkets & Other Grocery Stores
53100	Safeway	Grocers-Retail	44511003	Supermarkets & Other Grocery Stores
54100	Trading Co Store	Grocers-Retail	44511003	Supermarkets & Other Grocery Stores

Appendix B: List of all Social Gathering Facilities within the Study Area

DesID	OWNER_NAME	MUNICIPAL	NAICS_CODE	PROP_USE
1200	LATAH COUNTRY BIBLE CHURCH	Latah	813110	Churches
2200	1ST PRES CHURCH FAIRFIELD	Fairfield	813110	Churches
3200	ZION LUTHERAN CHURCH 108	Fairfield	813110	Churches
4200	UPPER COLUMBIA MISSION	Fairfield	813110	Churches
5200	UPPER COLUMBIA MISSION	Fairfield	813110	Churches
6200	SPANGLE CHRISTIAN CHURCH	Spangle	813110	Churches
7200	SPANGLE CHRISTIAN CHURCH	Spangle	813110	Churches
8200	ROCKFORD UNITED METHODIST	Rockford	813110	Churches
9200	ROCKFORD UNITED METHODIST 162	Rockford	813110	Churches
10200	ST JOSEPH CATHOLIC PARISH-ROCKFORD	Rockford	813110	Churches
11200	UNITED METHODIST CHURCH	Cheney	813110	Churches
12200	UNITED METHODIST CHURCH	Cheney	813110	Churches
13200	EPISCOPL DIOCESE	Cheney	813110	Churches
14200	EMAN LUTH CHENEY	Cheney	813110	Churches
15200	EMAN LUTH CHENEY	Cheney	813110	Churches
16200	EMAN LUTH CHENEY	Cheney	813110	Churches
17200	CHENEY COMMUNITY CHURCH	Cheney	813110	Churches
18200	INTL CHURCH FOURSQUARE GOSPEL	Cheney	813110	Churches
19200	UNITED CHURCH OF CHRIST	Cheney	813110	Churches
20200	ST ROSE OF LIMA CATHOLIC PARISH-CHENEY	Cheney	813110	Churches
21200	ST ROSE OF LIMA CATHOLIC PARISH-CHENEY	Cheney	813110	Churches
22200	GREENACRES BAPTIST CHURCH	Cheney	813110	Churches
23200	CHURCH OF LATTER DAY SAINTS	Cheney	813110	Churches
24200	AMAZING GRACE FELLOWSHIP	Cheney	813110	Churches
25200	CHENEY CHURCH OF NAZARENE	Cheney	813110	Churches

Appendix B cont'd

DesID	OWNER_NAME	MUNICIPAL	NAICS_CODE	PROP_USE
26200	CHENEY CONG OF JEHOVAHS WITT	County	813110	Churches
27200	WESTSIDE CHAPEL	County	813110	Churches
28200	VALLEYFORD COMMUNITY CHURCH	County	813110	Churches
29200	MARSHALL COMMUNITY CHURCH	County	813110	Churches
30200	ST JOHNS LUTHERAN CHURCH	Medical Lake	813110	Churches
31200	ASSEMBLY OF GOD	Medical Lake	813110	Churches
32200	ST ANNE CATHOLIC PARISH-MEDICAL LAKE	Medical Lake	813110	Churches
33200	ST ANNE CATHOLIC PARISH-MEDICAL LAKE	Medical Lake	813110	Churches
34200	MEDICAL LAKE COMMUNITY CHURCH	Medical Lake	813110	Churches
35200	BISHOP OF CHURCH LDS	County	813110	Churches
36200	MARY OF ANGELS CONVENT	County	813110	Churches
37200	MARY OF ANGELS CONVENT	County	813110	Churches
38200	ST ROSE CONVENT	County	813110	Churches
39200	IMMACULATE HEART RETREAT CNTR	County	813110	Churches
40200	WINDSOR BAPTIST	County	813110	Churches
41200	MORAN METHODIST	County	813110	Churches
42200	ST STEPHENS EPIS	Spokane	813110	Churches
43200	ST STEPHENS EPISOCPAL	Spokane	813110	Churches
44200	ST STEPHENS EPISCOPAL	Spokane	813110	Churches
45200	ST JOHN'S EVANGELICAL LUTHERN CHURCH	Spokane	813110	Churches
46200	THE KINGS COMM	County	813110	Churches
47200	SOUTH PERRY CONG. JEHOVAH'S WITNESS	County	813110	Churches
48200	LIVING WORD CHRISTIAN CENTER	County	813110	Churches
49200	LIVING WORD CHRISTIAN CENTER	County	813110	Churches
50200	UPPER COLUMBIA MISSION	County	813110	Churches
51200	LIVING WORD CHRISTIAN CENTER	County	813110	Churches
52200	PARK HEIGHTS BAPTIST CHURCH	County	813110	Churches
54200	CRESTLINE BAPTIST CHURCH	Spokane	813110	Churches

APPENDIX

Appendix B cont'd

DesID	OWNER_NAME	MUNICIPAL	NAICS_CODE	PROP_USE
55200	BEAUTIFUL SAVIOR CHURCH	Spokane	813110	Churches
57200	UPPER COLUMBIA MISSION	County	813110	Churches
58200	CHURCH OF LATTER DAY SAINTS	Spokane	813110	Churches
59200	HAMBLEN PARK PRESBYTERIAN CHUR	Spokane	813110	Churches
60200	UPPER COLUMBIA MISSION	County	813110	Churches
61200	CHESTER COMMUNITY CHURCH	Spokane Valley	813110	Churches
62200	CHESTER COMMUNITY CHURCH	Spokane Valley	813110	Churches
63200	OUR LADY OF FATIMA CATHOLIC PARISH	Spokane	813110	Churches
64200	MANITO UNITED METHODIST CHURCH	Spokane	813110	Churches
66200	OUR LADY OF FATIMA CATHOLIC PARISH	Spokane	813110	Churches
67200	THE CARMEL OF THE HOLY TRINITY	Spokane Valley	813110	Churches
68200	MANITO UNITED METHODIST CHURCH	Spokane	813110	Churches
69200	MANITO UNITED METHODIST CHURCH	Spokane	813110	Churches
70200	OUR LADY OF FATIMA CATHOLIC PARISH	Spokane	813110	Churches
71200	CHURCH OF JESUS CHRIST OF LDS	Spokane Valley	813110	Churches
72200	LIVING WORD CHRISTIAN CENTER	Spokane	813110	Churches
74200	OUR LADY OF FATIMA CATHOLIC PARISH	Spokane	813110	Churches
75200	CHURCH OF CHRIST LDS	County	813110	Churches
76200	1ST CH OF RELIGIOUS SCIENCE	Spokane	813110	Churches
77200	UNITY CHURCH OF TRUTH	Spokane	813110	Churches
78200	TEMPLE BETH SHALOM	Spokane	813110	Churches
79200	MANITO PRESBYTERIAN CHURCH	Spokane	813110	Churches
80200	SOUTH HILL BIBLE	Spokane	813110	Churches
81200	SOUTH HILL BIBLE	Spokane	813110	Churches
82200	LATTER DAY SAINTS	Spokane	813110	Churches
83200	LATTER DAY SAINTS	Spokane	813110	Churches
84200	LATTER DAY SAINTS	Spokane	813110	Churches
85200	LATTER DAY SAINTS	Spokane	813110	Churches

Appendix B cont'd

DesID	OWNER_NAME	MUNICIPAL	NAICS_CODE	PROP_USE
86200	REDEEMER LUTHERAN CHURCH	Spokane Valley	813110	Churches
87200	LINCOLN HTS CONGREGATIONAL CHU	Spokane	813110	Churches
88200	LINCOLN HTS CONGREGATIONAL CHU	Spokane	813110	Churches
90200	SUNSET HILL CONGREGATION OF	County	813110	Churches
91200	GRACE SLAVIC BAPTIST CHURCH	County	813110	Churches
93200	SOUTHSIDE CHRISTIAN CHURCH	Spokane	813110	Churches
94200	BETHLEHEM LUTHERAN CHURCH	Spokane	813110	Churches
97200	CHRIST HOLY SANC	Spokane	813110	Churches
98200	SOUTHSIDE ASSEMBLY OF GOD	Spokane	813110	Churches
99200	ST MARKS EVANGELICAL LUTHERAN CHURCH	Spokane	813110	Churches
100200	ST MARKS EVANGELICAL LUTHERAN CHURCH	Spokane	813110	Churches
101200	ST MARKS EVANGELICAL LUTHERAN CHURCH	Spokane	813110	Churches
102200	GARDEN SPRINGS CHURCH OF GOD	County	813110	Churches
103200	ST MARKS EVANGELICAL LUTHERAN CHURCH	Spokane	813110	Churches
104200	ST MARKS EVANGELICAL LUTHERAN CHURCH	Spokane	813110	Churches
105200	GARDEN SPRINGS CHURCH OF GOD	County	813110	Churches
106200	MCCLAIN, ELLEN G	County	813110	Churches
107200	VALLEY BIBLE CHURCH OF SPO	County	813110	Churches
108200	PACIFIC REGION OF OPEN BIBLE Churches	Spokane	813110	Churches
109200	ST AUGUSTINE CATHOLIC PARISH-SPOKANE	Spokane	813110	Churches
110200	TANGEN PROPERTIES, LLC	Airway Heights	813110	Churches
111200	ST AUGUSTINE CATHOLIC PARISH-SPOKANE	Spokane	813110	Churches
112200	ST AUGUSTINE CATHOLIC PARISH-SPOKANE	Spokane	813110	Churches
113200	AIRWAY HTS 1ST BAPTIST CHURCH	Airway Heights	813110	Churches
114200	ST AUGUSTINE CATHOLIC PARISH-SPOKANE	Spokane	813110	Churches
115200	HOLY TRNITY LUTHERAN	Spokane Valley	813110	Churches
116200	ST PETER CATHOLIC PARISH-SPOKANE	Spokane	813110	Churches
117200	AIRWAY EVANGELICAL FREE CHURCH	Airway Heights	813110	Churches

APPENDIX

Appendix B cont'd

DesID	OWNER_NAME	MUNICIPAL	NAICS_CODE	PROP_USE
118200	ST PETER CATHOLIC PARISH-SPOKANE	Spokane	813110	Churches
119200	ST PETER CATHOLIC PARISH-SPOKANE	Spokane	813110	Churches
120200	ST PETER CATHOLIC PARISH-SPOKANE	Spokane	813110	Churches
121200	VALLEY FOURTH MEMORIAL CHURCH	Spokane Valley	813110	Churches
122200	FIRST CHURCH CHRISTIAN SCIENCE	Spokane	813110	Churches
123200	ONE SPOKANE	Spokane Valley	813110	Churches
124200	ONE SPOKANE	Spokane Valley	813110	Churches
128200	ST JOHN CATHEDRAL EPIS	Spokane	813110	Churches
132200	ST JOHN CATHEDRAL EPIS	Spokane	813110	Churches
133200	ST JOHN CATHEDRAL EPIS	Spokane	813110	Churches
134200	ST JOHN CATHEDRAL EPIS	Spokane	813110	Churches
137200	CHRIST COMMUNITY CHURCH	Spokane	813110	Churches
139200	SPO VALLEY CHURCH OF NAZARENE	Spokane Valley	813110	Churches
140200	LIBERTY PARK METHODIST CHURCH	Spokane	813110	Churches
141200	LIBERTY PARK METHODIST CHURCH	Spokane	813110	Churches
142200	SACRED HEART CATHOLIC PARISH	Spokane	813110	Churches
144200	ST MATTHEWS INSTITUTIONAL BAPTIST CHURCH	Spokane	813110	Churches
146200	BETHEL CHURCH OF NAZAVENE	Spokane	813110	Churches
147200	SPOKANE BUDDHIST	Spokane	813110	Churches
148200	BEYOND GRACE FELLOWSHIP	Spokane	813110	Churches
151200	BEYOND GRACE FELLOWSHIP	Spokane	813110	Churches
152200	BIBLE BAPTIST CHURCH OF SPOKANE	Spokane Valley	813110	Churches
153200	PLYMOUTH CONGREGATIONAL	Spokane	813110	Churches
154200	UPPER COL CORP	Spokane Valley	813110	Churches
155200	PLYMOUTH CONGREGATIONAL	Spokane	813110	Churches
156200	PENTECOSTAL MISSIONARY CHURCH	Spokane	813110	Churches
157200	PLYMOUTH CONGREGATIONAL	Spokane	813110	Churches
158200	CHRIST THE SAVIOR ANTIOCHIAN ORTHODOX CH	Spokane Valley	813110	Churches

Appendix B cont'd

DesID	OWNER_NAME	MUNICIPAL	NAICS_CODE	PROP_USE
159200	KIRSCHKE, STEVEN J & DONNA	Spokane	813110	Churches
160200	CH LATTER DAY ST	Spokane Valley	813110	Churches
161200	CHURCH OF CHRIST	Spokane	813110	Churches
162200	OPPORTUNITY CHRISTIAN FELLOWSHIP	Spokane Valley	813110	Churches
164200	CHURCH OF CHRIST	Spokane	813110	Churches
165200	OPPORTUNITY CHRISTIAN FELLOWSHIP	Spokane Valley	813110	Churches
166200	BETHEL AMERICAN CHURCH	Spokane	813110	Churches
167200	CENTRAL LUTH	Spokane	813110	Churches
168200	CENTRAL LUTH	Spokane	813110	Churches
170200	OPPORTUNITY CHRISTIAN FELLOWSHIP	Spokane Valley	813110	Churches
173200	REYES, ANGEL THOMAS & LOURDES M	Spokane	813110	Churches
174200	OPPORTUNITY CHRISTIAN FELLOWSHIP	Spokane Valley	813110	Churches
175200	SPOKANE VALLEY BAPTIST CHURCH	Spokane Valley	813110	Churches
177200	REYES, ANGEL THOMAS & LOURDES M	Spokane	813110	Churches
179200	CENTRAL LUTH	Spokane	813110	Churches
180200	HIGHLND PARK METHODIST	Spokane	813110	Churches
181200	WESTMINSTER CONG CHURCH	Spokane	813110	Churches
182200	EMMANUEL LUTHERAN CHURCH	Spokane	813110	Churches
183200	WESTMINSTER CONG CHURCH	Spokane	813110	Churches
184200	FIRST PRESBYTERIAN CHURCH 181	Spokane	813110	Churches
185200	FIRST PRESBYTERIAN CHURCH	Spokane	813110	Churches
186200	FIRST PRESBYTERIAN CHURCH	Spokane	813110	Churches
187200	FIRST PRESBYTERIAN CHURCH 181	Spokane	813110	Churches
188200	FIRST PRESBYTERIAN CHURCH	Spokane	813110	Churches
189200	VALLEY VIEW BAPTIST	Spokane Valley	813110	Churches
190200	VALLEY VIEW BAPTIST	Spokane Valley	813110	Churches
191200	FIRST PRESBYTERIAN CHURCH 181	Spokane	813110	Churches
192200	MT ZION HOLINESS CHURCH	Spokane	813110	Churches

APPENDIX

Appendix B cont'd

DesID	OWNER_NAME	MUNICIPAL	NAICS_CODE	PROP_USE
193200	CENTRAL METHODIST CHURCH	Spokane	813110	Churches
195200	CENTRAL METHODIST CHURCH	Spokane	813110	Churches
196200	MT OLIVE BAPTIST	Spokane	813110	Churches
199200	ORCHARD CHRISTIAN FELLOWSHIP	Spokane	813110	Churches
201200	PILGRIM SLAVIC BAPTIST CHURCH	Spokane	813110	Churches
202200	CALVARY BAPTIST CHURCH	Spokane	813110	Churches
203200	CALVARY BAPTIST CHURCH	Spokane	813110	Churches
204200	NEW HOPE BAPTIST CHURCH	Spokane	813110	Churches
205200	NEW HOPE BAPTIST CHURCH	Spokane	813110	Churches
206200	GLAD TIDNGS ASSEMBLY OF GOD	Spokane	813110	Churches
207200	CALVARY BAPTIST CHURCH	Spokane	813110	Churches
209200	FIRST COVENANT SPOKANE	Spokane	813110	Churches
210200	FIRST COVENANT SPOKANE	Spokane	813110	Churches
214200	LATTER DAY SAINTS	Spokane Valley	813110	Churches
219200	STATE OF WASHINGTON DOT	Spokane	813110	Churches
220200	STATE OF WASHINGTON DOT	Spokane	813110	Churches
222200	PINES BAPTIST CHURCH	Spokane Valley	813110	Churches
223200	HEALING ROOMS MINISTRIES	Spokane	813110	Churches
224200	GOSEPL MEETING HALL	Spokane	813110	Churches
225200	GOOD SHEPHERD LUTHERAN CHURCH	Spokane Valley	813110	Churches
226200	MANZ, MICHAEL & PATRICIA	Spokane Valley	813110	Churches
227200	KOREAN PRESBYTERIAN CHURCH	Spokane	813110	Churches
228200	CHURCH OF GOD IN CHRIST CORP	Spokane	813110	Churches
229200	ALL NATIONS CHRISTIAN CENTER	Spokane	813110	Churches
230200	HEALING ROOMS MINISTRIES	Spokane	813110	Churches
232200	HEALING ROOMS MINISTRIES	Spokane	813110	Churches
233200	ST ANN CATHOLIC PARISH - SPOKANE	Spokane	813110	Churches
234200	EPISCOPAL CHURCH	Spokane Valley	813110	Churches

APPENDIX

Appendix B cont'd

DesID	OWNER_NAME	MUNICIPAL	NAICS_CODE	PROP_USE
236200	ST ANN CATHOLIC PARISH - SPOKANE	Spokane	813110	Churches
238200	FULL GOSPEL MISSION	Spokane	813110	Churches
240200	ST ANN CATHOLIC PARISH - SPOKANE	Spokane	813110	Churches
241200	CHURCH OF LATTER DAY SAINTS	Spokane Valley	813110	Churches
242200	CATHEDRAL OF OUR LADY OF LOURDES-SPOKANE	Spokane	813110	Churches
244200	CATHEDRAL OF OUR LADY OF LOURDES-SPOKANE	Spokane	813110	Churches
245200	THE NORTH AMERICAN ISLAMIC TRUST	Spokane Valley	813110	Churches
246200	LISAC'S OF WASHINGTON, INC	Spokane Valley	813110	Churches
247200	LIBERTY LAKE COMMUINTY CHURCH	County	813110	Churches
248200	ST MARY CATHOLIC PARISH-SPOKANE	Spokane Valley	813110	Churches
249200	LINCOLN HILLS, LLC	Spokane Valley	813110	Churches
250200	SPOKANE VAL UNITED METHODIST	Spokane Valley	813110	Churches
251200	SALEM LUTHERAN CHURCH	Spokane	813110	Churches
255200	DISHMAN FIRST BAPTIST CHURCH	Spokane Valley	813110	Churches
256200	CHRIST THE REDEEMER	Spokane	813110	Churches
261200	GREENACRES BAPTIST CHURCH	Spokane Valley	813110	Churches
263200	WASSON FAMILY, LLC	Spokane Valley	813110	Churches
264200	Library	Spokane Valley		Library
265200	ST JOSEPH CATHOLIC PARISH-SPOKANE	Spokane	813110	Churches
267200	ST JOHN VIANNEY CATHOLIC PARISH	Spokane Valley	813110	Churches
268200	HOLY TRNITY EPISCOPAL	Spokane	813110	Churches
269200	ST JOSEPH CATHOLIC PARISH-SPOKANE	Spokane	813110	Churches
270200	HOLY TRNITY EPISCOPAL	Spokane	813110	Churches
271200	HOLY TRNITY EPISCOPAL	Spokane	813110	Churches
272200	HOLY TRNITY EPISCOPAL	Spokane	813110	Churches
273200	ST JOHN VIANNEY CATHOLIC PARISH	Spokane Valley	813110	Churches
274200	INTERNATIONAL PENTECOSTAL CHURCH	Spokane	813110	Churches
275200	INTERNATIONAL PENTECOSTAL CHURCH	Spokane	813110	Churches

APPENDIX

Appendix B cont'd

DesID	OWNER_NAME	MUNICIPAL	NAICS_CODE	PROP_USE
276200	ST JOHN VIANNEY CATHOLIC PARISH	Spokane Valley	813110	Churches
277200	INTERNATIONAL PENTECOSTAL CHURCH	Spokane	813110	Churches
278200	INTERN'L CHURCH OF THE FOURSQUARE GOSPEL	Spokane	813110	Churches
279200	ST JOHN VIANNEY CATHOLIC PARISH	Spokane Valley	813110	Churches
280200	MORNING STAR BAPTIST	Spokane	813110	Churches
288200	WESTMINSTER PRES CHURCH	Spokane	813110	Churches
289200	MORNING STAR BAPTIST	Spokane	813110	Churches
290200	PENTECOSTAL CHURCH OF GOD	Spokane	813110	Churches
291200	GREENACRES TABERNACLE	Spokane Valley	813110	Churches
292200	ST ALOYSIUS CATHOLIC PARISH-SPOKANE	Spokane	813110	Churches
293200	ST ALOYSIUS CATHOLIC PARISH-SPOKANE	Spokane	813110	Churches
295200	HOYT, MICHAEL R	Spokane Valley	813110	Churches
296200	ST ALOYSIUS CATHOLIC PARISH-SPOKANE	Spokane	813110	Churches
297200	JESUS IS THE ANSWER CHURCH	Spokane	813110	Churches
298200	OPPORTUNITY CHRISTIAN CHURCH	Spokane Valley	813110	Churches
299200	LINCOLN CENTER SPOKANE LLC	Spokane	813110	Churches
301200	GONZAGA UNIVERSITY	Spokane	813110	Churches
302200	OPPORTUNITY CHRISTIAN CHURCH	Spokane Valley	813110	Churches
303200	LINCOLN CENTER SPOKANE LLC	Spokane	813110	Churches
304200	OPPORTUNITY BAPTIST CHURCH	Spokane Valley	813110	Churches
307200	COLUMBIA RIVER METH CHURCH	Spokane	813110	Churches
308200	COMMUNITY CONGREGATIONAL CHURCH OF VERA	Spokane Valley	813110	Churches
309200	COLUMBIA RIVER METH CHURCH	Spokane	813110	Churches
310200	LINCOLN CENTER SPOKANE LLC	Spokane	813110	Churches
312200	VALLEY ASSEMBLY OF GOD	Spokane Valley	813110	Churches
313200	CORP GONZAGA UNIVERSITY	Spokane	813110	Churches
317200	REORGANIZED CHURCH OF LATTER DAY SAINTS	Spokane Valley	813110	Churches
318200	COSTELLO, ROGER A & BRIDGET B	Spokane	813110	Churches

Appendix B cont'd

DesID	OWNER_NAME	MUNICIPAL	NAICS_CODE	PROP_USE
319200	CORNERSTONE	Spokane Valley	813110	Churches
320200	CHRIST LUTHERAN CHURCH	Spokane Valley	813110	Churches
321200	UNITARIAN CHURCH OF SPOKANE	Spokane	813110	Churches
322200	GETHSEMANE LUTHERAN	Spokane Valley	813110	Churches
323200	UNITARIAN CHURCH OF SPOKANE	Spokane	813110	Churches
324200	NEW HOPE CHRISTIAN CENTER	Spokane Valley	813110	Churches
325200	UPPER COLUMBIA CORP	Spokane	813110	Churches
326200	HOLY NAMES SISTERS	Spokane	813110	Churches
327200	CALVARY CHAPEL OF SPO VALLEY	Spokane Valley	813110	Churches
329200	WESLEYAN CHURCH	Spokane Valley	813110	Churches
330200	UPPER COLUMBIA MISSION	Spokane	813110	Churches
331200	UPPER COLUMBIA MISSION	Spokane	813110	Churches
332200	UPPER COLUMBIA MISSION	Spokane	813110	Churches
333200	VALLEY LANDMARK BAPTIST	Spokane Valley	813110	Churches
334200	DOMINICAN SIS WN	Spokane	813110	Churches
335200	VALLEY CHURCH OF CHRIST	Spokane Valley	813110	Churches
336200	HOPE LUTHERAN CHURCH	Spokane Valley	813110	Churches
337200	BEREAN BIBLE CHURCH SPOKANE	Spokane Valley	813110	Churches
338200	ST PAUL METHODIST	Spokane	813110	Churches
341200	MISSION AVENUE PRESBYTERIAN	Spokane	813110	Churches
342200	BETHLEHEM SLAVIC CHURCH	Spokane	813110	Churches
344200	BETHLEHEM SLAVIC CHURCH	Spokane	813110	Churches
345200	GREEK ORTHODOX	Spokane	813110	Churches
346200	INT CHURCH FOUR SQUARE GOSPEL	Spokane	813110	Churches
347200	SISTERS OF HOLY NAME	Spokane	813110	Churches
348200	INT CHURCH FOUR SQUARE GOSPEL	Spokane	813110	Churches
349200	FIRST ASSEMBLY OF GOD	Spokane	813110	Churches
350200	FIRST ASSEMBLY OF GOD	Spokane	813110	Churches

APPENDIX

Appendix B cont'd

DesID	OWNER_NAME	MUNICIPAL	NAICS_CODE	PROP_USE
351200	HOLY TEMPLE CHURCH OF GOD	Spokane	813110	Churches
353200	FIRST ASSEMBLY OF GOD	Spokane	813110	Churches
354200	CORP. BISHOP OF CHURCH JESUS	Spokane	813110	Churches
359200	GREEENACRES JEHOVAHS	Spokane Valley	813110	Churches
360200	CORP. BISHOP OF CHURCH JESUS	Spokane	813110	Churches
361200	GREENACRES JEHOVAHS	Spokane Valley	813110	Churches
367200	CORP. BISHOP OF CHURCH JESUS	Spokane	813110	Churches
368200	FOURTH MEMORIAL CHURCH	Spokane	813110	Churches
369200	SISTERS OF HOLY NAME	Spokane	813110	Churches
371200	CENTRAL BAPTIST CHURCH	Spokane	813110	Churches
376200	SISTERS OF HOLY NAME	Spokane	813110	Churches
382200	FOURTH MEMORIAL CHURCH	Spokane	813110	Churches
383200	KNOX PRESBYTERIAN CHURCH	Spokane	813110	Churches
390200	FOURTH MEMORIAL CHURCH	Spokane	813110	Churches
398200	DOMINICAN SIS WN	Spokane	813110	Churches
399200	SIKH TEMPLE OF SPOKANE	Spokane Valley	813110	Churches
400200	GRACE CHRISTIAN FELLOWSHIP	Spokane	813110	Churches
401200	ST ANTHONY CATHOLIC PARISH-SPOKANE	Spokane	813110	Churches
402200	ST ANTHONY CATHOLIC PARISH-SPOKANE	Spokane	813110	Churches
404200	L.D.S. CHURCH	Liberty Lake	813110	Churches
405200	GRACE CHRISTIAN FELLOWSHIP	Spokane	813110	Churches
406200	UNIV CON OF JEHOV WITNESSES	Spokane Valley	813110	Churches
407200	RIVERVIEW ASSOCIATION	Spokane	813110	Churches
408200	SLAVIC CHRISTIAN EVANGELICAL CHURCH	Spokane	813110	Churches
409200	SLAVIC CHRISTIAN EVANGELICAL CHURCH	Spokane	813110	Churches
410200	GREENACRES CHRISTIAN CHURCH	Spokane Valley	813110	Churches
411200	BROOKSIDE PROPERTY DEVELOPMENT, LLC	Spokane	813110	Churches
412200	ST ANDREW EPIS	Spokane	813110	Churches

Appendix B cont'd

DesID	OWNER_NAME	MUNICIPAL	NAICS_CODE	PROP_USE
413200	ST ANDREW EPIS	Spokane	813110	Churches
414200	ST ANDREW EPIS	Spokane	813110	Churches
415200	CHURCH OF THE LIVING GOD INTNL	Spokane	813110	Churches
416200	MARY QUEEN CATHOLIC PARISH-SPOKANE	Spokane	813110	Churches
417200	EMMANUEL PRESBYTERIAN CHURCH	Spokane	813110	Churches
418200	EMMANUEL PRESBYTERIAN CHURCH	Spokane	813110	Churches
419200	MARY QUEEN CATHOLIC PARISH-SPOKANE	Spokane	813110	Churches
420200	ST GREGORIOS ORTHODOX MISSION SOCIETY	Spokane	813110	Churches
421200	MINNEHAHA COVN CHURCH	Spokane	813110	Churches
422200	MINNEHAHA COVN CHURCH	Spokane	813110	Churches
423200	ST PASCHAL CATHOLIC PARISH-SPOKANE	Spokane Valley	813110	Churches
424200	PILGRIM LUTHERAN	Spokane	813110	Churches
425200	HILLYARD JEHOVAHS	Spokane	813110	Churches
427200	PILGRIM LUTHERAN	Spokane	813110	Churches
428200	CHURCH OF GOD SEVENTH DAY	Spokane	813110	Churches
429200	CHURCH OF GOD SEVENTH DAY	Spokane	813110	Churches
430200	VICTORY FAITH FELLOWSHIP	Spokane Valley	813110	Churches
431200	MACKAY, DEBRA A & MICHAEL J	Spokane Valley	813110	Churches
432200	FAITH BAPTIST CHURCH OF SPOKANE	Spokane	813110	Churches
433200	SHERMAN HILLS STEWARDSHIP MINISTRIES	Millwood	813110	Churches
434200	CHURCH IN SPOKANE	Spokane	813110	Churches
435200	FIRST FREE METHODIST CHURCH	Spokane	813110	Churches
436200	FAITH BIBLE CHURCH	Spokane	813110	Churches
437200	FULL GOSPEL TABERNACLE	Spokane	813110	Churches
438200	FULL GOSPEL TABERNACLE	Spokane	813110	Churches
439200	FAITH BIBLE CHURCH	Spokane	813110	Churches
440200	TRINITY UNITED METHODIST	Spokane	813110	Churches
441200	FAITH BIBLE CHURCH	Spokane	813110	Churches

APPENDIX

Appendix B cont'd

DesID	OWNER_NAME	MUNICIPAL	NAICS_CODE	PROP_USE
442200	TRINITY UNITED METHODIST	Spokane	813110	Churches
443200	INTERNATIONAL CHURCH FOURSQUARE GOSPEL	Spokane Valley	813110	Churches
445200	MILLWOOD PRESBYTERIAN	Millwood	813110	Churches
448200	CHRIST CHURCH SPOKANE	Spokane	813110	Churches
449200	CHRIST CHURCH SPOKANE	Spokane	813110	Churches
450200	CHURCH OF NAZARENE	Spokane Valley	813110	Churches
451200	PIPKIN, BRADLEY J & ELIZABETH M	Spokane Valley	813110	Churches
452200	SOUTH HILL BAPTIST CHURCH	Spokane Valley	813110	Churches
454200	ST FRANCIS XAVIER CATHOLIC PARISH	Spokane	813110	Churches
455200	C & M ALLIANCE	Spokane	813110	Churches
456200	ST FRANCIS XAVIER CATHOLIC PARISH	Spokane	813110	Churches
457200	THE BRIDGE	Spokane	813110	Churches
458200	AUDUBON PARK METHODIST	Spokane	813110	Churches
459200	THE BRIDGE	Spokane	813110	Churches
460200	NORTHWEST JEHOVAHS	Spokane	813110	Churches
461200	ST FRANCIS XAVIER CATHOLIC PARISH	Spokane	813110	Churches
462200	CROSSWALK COMMUNITY CHURCH	Spokane	813110	Churches
463200	NATIONAL ORG OF THE NEW APOST CHURCH	Spokane	813110	Churches
465200	SECOND CHURCH OF CHRIST SCIENTIST	Spokane	813110	Churches
466200	FOWLER UNITED METHODIST	Spokane	813110	Churches
467200	OUR LADY OF GUADALUPE CHURCH	Spokane	813110	Churches
468200	FOWLER UNITED METHODIST	Spokane	813110	Churches
469200	OUR LADY OF GUADALUPE CHURCH	Spokane	813110	Churches
470200	INTERNATIONAL CHURCH OF THE	Spokane	813110	Churches
471200	INTERNATIONAL CHURCH OF THE	Spokane	813110	Churches
473200	TRENTWOOD CONG OF JEHOVAHS WITNESSES	Spokane Valley	813110	Churches
474200	MESSIAH LUTHERAN	Spokane	813110	Churches
475200	MONROE PARK GOSPEL	Spokane	813110	Churches

APPENDIX

Appendix B cont'd

DesID	OWNER_NAME	MUNICIPAL	NAICS_CODE	PROP_USE
476200	MONROE PARK GOSPEL	Spokane	813110	Churches
477200	MESSIAH LUTHERAN	Spokane	813110	Churches
478200	CORNERSTONE COMMUNITY CHURCH NW	Spokane	813110	Churches
479200	CORNERSTONE COMMUNITY CHURCH NW	Spokane	813110	Churches
480200	CHRIST GOSPEL FELLOWSHIP	Spokane	813110	Churches
481200	COMMUNITY BIBLE CHAPEL	Spokane	813110	Churches
482200	COMMUNITY BIBLE CHAPEL	Spokane	813110	Churches
483200	COVENANT CHURCH & COVENANT CHRISTIAN SCH	Spokane	813110	Churches
484200	ST CHARLES CATHOLIC PARISH-SPOKANE	Spokane	813110	Churches
485200	ST CHARLES CATHOLIC PARISH-SPOKANE	Spokane	813110	Churches
486200	COVENANT CHURCH & COVENANT CHRISTIAN SCH	Spokane	813110	Churches
487200	BETHEL MISSIONARY BAPTIST CHURCH	County	813110	Churches
488200	LIDGERWOOD PRESBYTERIAN CHURCH	Spokane	813110	Churches
489200	LIDGERWOOD PRESBYTERIAN CHURCH	Spokane	813110	Churches
490200	NAZARENE CHURCH	County	813110	Churches
491200	NAZARENE CHURCH	County	813110	Churches
495200	NAZARENE CHURCH	County	813110	Churches
496200	NORTH HILL CHRISTIAN CHURCH	Spokane	813110	Churches
499200	CRESTLINE NAZARENE	Spokane	813110	Churches
500200	DRISCOLL BLV BAPTIST CHURCH	Spokane	813110	Churches
501200	TRINITY LUTHERAN CHURCH	Spokane	813110	Churches
502200	SUNRISE CHURCH OF CHRIST	Spokane	813110	Churches
503200	ZION FAITH ASSEMBLY	Spokane	813110	Churches
504200	ST PETERS LUTHERAN	Spokane	813110	Churches
505200	ST PETERS LUTHERAN	Spokane	813110	Churches
506200	REFRESING SPRING CH OF GOD	Spokane	813110	Churches
507200	REFRESING SPRING CH OF GOD	Spokane	813110	Churches
508200	FIRST EVANGEL FREE CHURCH	Spokane	813110	Churches

APPENDIX

Appendix B cont'd

DesID	OWNER_NAME	MUNICIPAL	NAICS_CODE	PROP_USE
509200	FIRST EVANGEL FREE CHURCH	Spokane	813110	Churches
510200	BYZN BISHOP VAN NUYS,INC SOLE	Spokane Valley	813110	Churches
511200	BYZN BISHOP VAN NUYS,INC SOLE	Spokane Valley	813110	Churches
512200	HILLYARD BAPTIST	Spokane	813110	Churches
514200	NEW TESTAMENT CHRISTIAN CHURCH	Spokane	813110	Churches
518200	IMMANUEL BAPT	Spokane	813110	Churches
519200	IMMANUEL BAPT	Spokane	813110	Churches
520200	PEACE LUTHERAN CHURCH	County	813110	Churches
521200	ST NICHOLAS ANTIOCHIAN CHURCH	Spokane	813110	Churches
523200	RESURRECTION CHURCH	Spokane	813110	Churches
524200	HILLYARD ASSEMBLY OF GOD	Spokane	813110	Churches
525200	ST NICHOLAS ANTIOCHIAN CHURCH	Spokane	813110	Churches
526200	CATHOLIC CEMETERIES OF SPOK	County	813110	Churches
527200	COMMUNITY CHURCH OF BIBLE	Spokane Valley	813110	Churches
529200	ST PATRICK CATHOLIC PARISH-SPOKANE	Spokane	813110	Churches
532200	SPOKANE CHINESE BAPTIST CHURCH	Spokane	813110	Churches
533200	NEW BEGINNINGS OPEN BIBLE CHURCH	Spokane	813110	Churches
534200	ST PATRICK CATHOLIC PARISH-SPOKANE	Spokane	813110	Churches
535200	INTERNATIONAL SANGHA BHIKSU	Spokane	813110	Churches
536200	CENTER OF LIFE SLAVIC CHURCH	Spokane Valley	813110	Churches
537200	UPPER COLUMBIA CORPORATION	County	813110	Churches
538200	EAST VALLEY BAPTIST CHURCH	Spokane Valley	813110	Churches
539200	GLORIA DEI LUTHERAN CHURCH	Spokane	813110	Churches
540200	GLORIA DEI LUTHERAN CHURCH	Spokane	813110	Churches
541200	MORNING STAR BAPTIST CHURCH OF SPOKANE	Spokane	813110	Churches
543200	LINCOLN HEIGHTS CHURCH OF GOD	Spokane	813110	Churches
544200	CATHOLIC CEMETERIES OF SPOKANE	County	813110	Churches
545200	SHADLE PARK PRESBYTERIAN	Spokane	813110	Churches

Appendix B cont'd

DesID	OWNER_NAME	MUNICIPAL	NAICS_CODE	PROP_USE
547200	SHADLE PARK PRESBYTERIAN	Spokane	813110	Churches
549200	FREE METHODIST CHURCH OF WA	County	813110	Churches
550200	ST JOSEPH CATHOLIC PARISH-OTIS ORCHARDS	County	813110	Churches
552200	LDS CHURCH TAX ADMIN.	County	813110	Churches
553200	RIDGEVIEW ASSEMBLY OF GOD	Spokane	813110	Churches
554200	FIRST CHURCH OF CHRIST SPOKANE	Spokane	813110	Churches
555200	FIRST CHURCH OF CHRIST SPOKANE	Spokane	813110	Churches
556200	LDS CHURCH TAX ADMIN.	County	813110	Churches
557200	RIDGEVIEW ASSEMBLY OF GOD	Spokane	813110	Churches
560200	OTIS ORCHRDS COMM CHURCH	County	813110	Churches
561200	MARANTHA CHURCH OF SPOKANE	County	813110	Churches
562200	OTIS ORCHRDS COMM CHURCH	County	813110	Churches
563200	STARR ROAD BAPTIST	County	813110	Churches
564200	STARR ROAD BAPTIST	County	813110	Churches
565200	IGLESIA NI CHRISTO (CHURCH OF CHRIST)	Spokane	813110	Churches
566200	CHURCH OF CHRIST LDS	Spokane Valley	813110	Churches
567200	FIRST FRIENDS CHURCH	Spokane	813110	Churches
568200	NORTH ADDISON BAPTIST	Spokane	813110	Churches
569200	EAST VAL PRESBYTERIAN CHURCH	County	813110	Churches
570200	NEW HOPE CHRISTIAN REFORMED CHURCH	Spokane	813110	Churches
571200	WEST SIDE CHURCH NAZARENE	Spokane	813110	Churches
572200	CHURCH OF CHRIST - LDS	Spokane	813110	Churches
573200	ROCK OF AGES	Spokane	813110	Churches
574200	LIGHT OF GOSPEL BAPTIST CHRUCH	County	813110	Churches
575200	UPPER COLUMBIA MISSION	County	813110	Churches
576200	LIGHT OF GOSPEL BAPTIST CHRUCH	County	813110	Churches
577200	ROCK OF AGES	Spokane	813110	Churches
578200	PLAZA 1 LIMITED, LLC	Spokane	813110	Churches

APPENDIX

Appendix B cont'd

DesID	OWNER_NAME	MUNICIPAL	NAICS_CODE	PROP_USE
579200	NORTHWOOD PRESBYTERIAN	County	813110	Churches
580200	PRINCE OF PEACE LUTHERAN CHUR	Spokane	813110	Churches
581200	ST MATTHEWS LUTHERAN	County	813110	Churches
582200	ASSUMPTION OF THE BLESSED VIRGIN CATHOLI	Spokane	813110	Churches
584200	HILLYARD CHRISTIAN CHURCH	Spokane	813110	Churches
585200	CHURCH OF LATTER DAY SAINTS	Spokane	813110	Churches
586200	EXCEL YOUTH CENTER	Spokane	813110	Churches
587200	ALPHA CHAPEL CHRISTIAN FAMILY	County	813110	Churches
588200	HOLY CROSS LUTHERAN	Spokane	813110	Churches
589200	REGAL STREET BAPTIST	County	813110	Churches
590200	SPOKANE CHRISTIAN CENTER	County	813110	Churches
591200	NEW HORIZONS COMMUNITY CHURCH	County	813110	Churches
592200	SHILOH HILLS FELLOWSHIP	Spokane	813110	Churches
593200	WORD OF LIFE COMM CHURCH	County	813110	Churches
594200	FIRST CHURCH OF THE OPEN BIBLE	County	813110	Churches
595200	UNITY CHURCH OF SPOKANE	County	813110	Churches
596200	PRINCE OF PEACE LUTH. CHURCH	Spokane	813110	Churches
597200	COUNTRY HOMES CHRISTIAN CHURCH	County	813110	Churches
598200	INDIAN TRAIL COMMUNITY CHURCH	Spokane	813110	Churches
599200	MT VIEW ASSEMBLY OF GOD	Spokane	813110	Churches
600200	1ST CHURCH OF THE NAZARENE	County	813110	Churches
601200	UPPER COLUMBIA MISSION	County	813110	Churches
602200	SPOKANE SLAVIC BAPTIST CHURCH	County	813110	Churches
603200	SPOKANE BIBLE CHURCH	County	813110	Churches
604200	LATTER DAY SAINTS CHURCH	Spokane	813110	Churches
605200	FIVE MILE COMMUNITY CHURCH	County	813110	Churches
606200	COUNTRY HOMES JEHOVAHS WIT	County	813110	Churches
607200	CHUBENKO, NIKOLAY & TATYANA	County	813110	Churches

Appendix B cont'd

DesID	OWNER_NAME	MUNICIPAL	NAICS_CODE	PROP_USE
608200	ST LUKE LUTHERAN CHURCH	County	813110	Churches
609200	CHURCH OF LATTER DAY SAINTS	County	813110	Churches
610200	FOOTHILLS COMMUNITY CHURCH	County	813110	Churches
611200	FIRST CHURCH OPEN BIBLE	County	813110	Churches
612200	NW BIBLE CHURCH	County	813110	Churches
613200	MEAD UNITED METHODIST CHURCH	County	813110	Churches
614200	FAIRWOOD BAPT CH	County	813110	Churches
615200	NORTHVIEW BIBLE CHURCH	County	813110	Churches
616200	CHRISTIAN LIFE CHURCH	County	813110	Churches
617200	COMMUNITY BIBLE CHAPEL	County	813110	Churches
618200	COVENANT UNITED METHODIST CHURCH	County	813110	Churches
619200	CHURCH OF CHRIST - COLBERT	County	813110	Churches
620200	CHURCH OF JESUS CHRIST LDS	County	813110	Churches
621200	CROSSOVER CHURCH	County	813110	Churches
622200	ST JOSEPH CATHOLIC PARISH-COLBERT	County	813110	Churches
623200	COLBERT PRESBYTERIAN CHURCH	County	813110	Churches
624200	COLBERT CHAPEL	County	813110	Churches
625200	COLBERT CHAPEL	County	813110	Churches
626200	GREEN BLUFF UNITED METHODIST	County	813110	Churches
627200	WILD ROSE UN METHODIST CHURCH	County	813110	Churches
628200	CHATTAROY COMMUNITY CHURCH	County	813110	Churches
629200	GRACE CHURCH OF DEER PARK	County	813110	Churches
630200	KINGDOM HALL OF JEHOVAH'S WITNESS	Deer Park	813110	Churches
631200	FAITH LUTHERAN CH OF DEER PARK	Deer Park	813110	Churches
632200	DEER PARK CHURCH OF CHRIST	County	813110	Churches
633200	ZION EVANGELICAL LUTH CHURCH	Deer Park	813110	Churches
634200	PACIFIC NORTHWEST ASSOC THE CHURCH GOD	Deer Park	813110	Churches
635200	OPEN DOOR CONG	Deer Park	813110	Churches

Appendix B cont'd

DesID	OWNER_NAME	MUNICIPAL	NAICS_CODE	PROP_USE
636200	ST MARY PRESENTATION CATHOLIC PARISH	Deer Park	813110	Churches
637200	ST MARY PRESENTATION CATHOLIC PARISH	Deer Park	813110	Churches
638200	DEER PARK CHURCH OF CHRIST	County	813110	Churches
639200	CHURCH OF CHRIST LDS	County	813110	Churches
640200	LAESTADIAN LUTHERN CONGREGATION	County	813110	Churches
641200	FOOTHLLS MISSION CHURCH	County	813110	Churches
642200	COUNTRY CHURCH OF THE OPEN BIB	County	813110	Churches
643200	ELK CONG CHURCH	County	813110	Churches
644200	COUNTRY CHURCH OF THE OPEN BIB	County	813110	Churches
645200	PICKENS, CARL W & ETAL	Latah		Public Assembly
646200	JERRYS FARMING SUPPLY, LP	County		Public Assembly
647200	WAVERLY MASONIC TEMPLE	Waverly		Public Assembly
648200	LASZ LIVING TRUST	Waverly		Public Assembly
649200	TYLER GRANGE #610	County		Public Assembly
650200	CAMPBELL ETL, J J	Spangle		Public Assembly
651200	SPANGLE SER CLUB	Spangle		Public Assembly
652200	MCINTOSH GRANGE	Rockford		Public Assembly
653200	AMERICAN LEGION #72	Cheney		Public Assembly
654200	MCMILLAN, DAVID L & VICKIE L	County		Public Assembly
655200	DEVILBISS, JAMES A	County		Public Assembly
656200	MEDICAL LAKE FOOD BANK	Medical Lake		Public Assembly
657200	JENSEN MENORIAL YOUTH RANCH	Medical Lake		Public Assembly
658200	ESPANOLA GRANGE	County		Public Assembly
659200	MORAN PR GRNG161	County		Public Assembly
660200	WINDSOR GR 980	County		Public Assembly
661200	WEST DEEP CREEK GRANGE	County		Public Assembly
662200	GREENACRES GRNGE	County		Public Assembly
663200	AIRWAY HEIGHTS	Airway Heights		Public Assembly

Appendix B cont'd

DesID	OWNER_NAME	MUNICIPAL	NAICS_CODE	PROP_USE
664200	WOMANS CLUB OF SPOKANE	Spokane		Public Assembly
666200	ALANO CLUB INC	Spokane		Public Assembly
668200	SHEVCHENKO, KLAVDIYA & MILKHAIL	Spokane		Public Assembly
669200	GESELLSCHAFT, D	Spokane		Public Assembly
670200	SPOKANE PARTNERS LLC	Spokane		Public Assembly
671200	SPOKANE PARTNERS LLC	Spokane		Public Assembly
690200	NEW FOX THEATER LLC	Spokane		Public Assembly
691200	GOOD SAMARITAN	Spokane Valley		Public Assembly
692200	THE NEW MET, LLC	Spokane		Public Assembly
693200	GENTILE/MAISEL/DUNCAN	Spokane		Public Assembly
694200	VFW 1435	Spokane Valley		Public Assembly
697200	MASONIC TEMPLE	Spokane		Public Assembly
699200	MASONIC TEMPLE	Spokane		Public Assembly
700200	SPOKANE CITY CLUB	Spokane		Public Assembly
701200	SPOKANE, CITY OF	Spokane		Public Assembly
702200	SECURE-IT SELF STORAGE, LLC	Spokane Valley		Public Assembly
703200	SPOKANE PUBLIC FACILITIES DISTRICT	Spokane		Public Assembly
704200	SPOKANE PUBLIC FACILITIES DISTRICT	Spokane		Public Assembly
706200	SP VALLEY F O E	Spokane Valley		Public Assembly
707200	SPOKANE PUBLIC FACILITIES DISTRICT	Spokane		Public Assembly
713200	SPOKANE PUBLIC FACILITIES DISTRICT	Spokane		Public Assembly
714200	SPOKANE PUBLIC FACILITIES DISTRICT	Spokane		Public Assembly
715200	SPOKANE PUBLIC FACILITIES DISTRICT	Spokane		Public Assembly
716200	SPOKANE CIVIC THEATRE	Spokane		Public Assembly
718200	SPOKANE CIVIC THEATRE	Spokane		Public Assembly
719200	SPOKANE PUBLIC FACILITIES DISTRICT	Spokane		Public Assembly
720200	LIPES RENTALS LLC	Spokane		Public Assembly
723200	SPOKANE PUBLIC FACILITIES DISTRICT	Spokane		Public Assembly

APPENDIX

Appendix B cont'd

DesID	OWNER_NAME	MUNICIPAL	NAICS_CODE	PROP_USE
724200	SPOKANE PUBLIC FACILITIES DISTRICT	Spokane		Public Assembly
728200	SPOKANE PUBLIC FACILITIES DISTRICT	Spokane		Public Assembly
730200	KNIGHTS OF COLMB	Spokane		Public Assembly
731200	SENIOR CITIZEN COUNCIL	Spokane		Public Assembly
732200	SENIOR CITIZEN COUNCIL	Spokane		Public Assembly
733200	SENIOR CITIZEN COUNCIL	Spokane		Public Assembly
734200	INDEPENDENT ORDER ODD FELLOW	Spokane		Public Assembly
735200	INDEPENDENT ORDER ODD FELLOW	Spokane		Public Assembly
736200	VETERANS FOREIGN AFFAIRS	Spokane		Public Assembly
737200	PINES 1212 LLC	Spokane Valley		Public Assembly
738200	E SPOKANE GRANGE 148	Spokane Valley		Public Assembly
739200	GREATER SPOKANE ELKS LODGE #228	Spokane Valley		Public Assembly
740200	NW BLVD TEMPLE	Spokane		Public Assembly
741200	LIFE RIVER FELLOWSHIP	Spokane Valley		Public Assembly
742200	CONCORDIA TEMPLE BUILDING	Millwood		Public Assembly
744200	SPOKANE, CITY OF	Spokane		Public Assembly
748200	FRITCHIE, KATHERINE M	Spokane		Public Assembly
750200	FRITCHIE, KATHERINE M	Spokane		Public Assembly
752200	NO HILL LODGE	Spokane		Public Assembly
755200	CANTU INVESTMENTS, LLC	Spokane		Public Assembly
759200	SNOWBIRD DIVISION, LLC	Spokane		Public Assembly
760200	CENTRAL GRG 831	County		Public Assembly
761200	SPRING HILL GRANGE #909	County		Public Assembly
762200	CALVARY CHAPEL OF SPOKANE INC	County		Public Assembly
763200	BEACON LODGE 91	County		Public Assembly
764200	WANDERMERE CO	County		Public Assembly
765200	SKAUG, EDWIN R	County		Public Assembly
766200	GREENBLUFF GRANGE 300	County		Public Assembly

Appendix B cont'd

DesID	OWNER_NAME	MUNICIPAL	NAICS_CODE	PROP_USE
767200	HALF MOON GRANGE	County		Public Assembly
768200	WILDROSE COMMUNITY CENTER	County		Public Assembly
769200	FRATERNAL ORDER OF EAGLES	Deer Park		Public Assembly
770200	FRATERNAL ORDER OF EAGLES	Deer Park		Public Assembly
771200	INLAND GRANGE	County		Public Assembly
772200	INLAND GRANGE	County		Public Assembly
773200	VFW 5924	County		Public Assembly
774200	Spokane Public Library	Spokane		Library
775200	SPOKANE PUBLIC LIBRARY	Spokane		Library
776200		Spokane		Library
777200	Spokane Public Library	Spokane		Library
778200	Moran Prairie Library	County		Library
779200	Argonne Library	County	813110	Churches
780200	OPPORTUNITY PRESBYTERIAN CH	Spokane Valley	813110	Churches
781200	Library	Spokane	813110	Churches
782200	North Spokane Library	County		Library
783200	Harriet Cheney Cowles Memorial Library	County		Library
784200	Spokane Public Library	Spokane		Library
785200	AIRWAY HEIGHTS	Airway Heights		Library
786200	Spokane County Library	Medical Lake		Library
787200	Spokane County Library	Cheney		Library

Appendix C: List of all Sports and Recreation Facilities within the Study Area

DesID	COMPANY_NA	PRIMARY_CI	NAICS_CODE	NAICS_DESC
1300	Curves	Airway Heights	71394011	Fitness & Recreational Sports Centers
2300	Antler Springs Golf Course	Chattaroy	71391002	Golf Courses & Country Clubs
3300	Creative Fulfillment	Cheney	71399002	All Other Amusement & Recreation Industries
4300	Curves	Cheney	71394011	Fitness & Recreational Sports Centers
5300	Evans Farms	Cheney	71399050	All Other Amusement & Recreation Industries
6300	Fairways Golf Course	Cheney	71391002	Golf Courses & Country Clubs
7300	Marshland Equestrian Ctr	Cheney	71399050	All Other Amusement & Recreation Industries
8300	Pro-Form Personal Training Inc.	Cheney	71394013	Fitness & Recreational Sports Centers
9300	Sante Fitness & Massage	Cheney	71394013	Fitness & Recreational Sports Centers
10300	Anytime Fitness	Deer Park	71394011	Fitness & Recreational Sports Centers
11300	Blue Haven Stables	Deer Park	71399050	All Other Amusement & Recreation Industries
12300	Curves	Deer Park	71394011	Fitness & Recreational Sports Centers
13300	Deer Park Riding School	Deer Park	71399050	All Other Amusement & Recreation Industries
14300	Diamond D Horse Boarding	Deer Park	71399050	All Other Amusement & Recreation Industries
15300	Farwest Farms LLC	Deer Park	71399050	All Other Amusement & Recreation Industries
16300	Northwest Trails	Deer Park	71399050	All Other Amusement & Recreation Industries
17300	Spur-the Moment Equestrian Ctr	Deer Park	71399050	All Other Amusement & Recreation Industries
18300	Fairchild Bowling Lanes Pizza	Fairchild Afb	71395001	Bowling Centers
19300	Affordable RV & Boat Storage	Spokane Valley	71393004	Marinas
20300	Indoor Golf LLC	Spokane Valley	71391002	Golf Courses & Country Clubs
21300	Anew Start Rejuvenation Ctr	Liberty Lake	71394011	Fitness & Recreational Sports Centers
22300	Anytime Fitness	Liberty Lake	71394011	Fitness & Recreational Sports Centers
23300	Curves	Liberty Lake	71394011	Fitness & Recreational Sports Centers
24300	Liberty Lake Athletic Club	Liberty Lake	71394011	Fitness & Recreational Sports Centers
25300	Liberty Lake Golf Course	Liberty Lake	71391002	Golf Courses & Country Clubs

Appendix C cont'd

DesID	COMPANY_NA	PRIMARY_CI	NAICS_CODE	NAICS_DESC
26300	Meadowwood Golf Course	Liberty Lake	71391002	Golf Courses & Country Clubs
27300	Trail Head Golf Course	Liberty Lake	71391002	Golf Courses & Country Clubs
28300	Inland Empire Paintball	Mead	71399031	All Other Amusement & Recreation Industries
29300	Bodyscapes	Medical Lake	71394013	Fitness & Recreational Sports Centers
30300	G & H Boat Storage Inc	Newman Lake	71393004	Marinas
31300	Landt Farms Sporting Clays	Nine Mile Fall	71399051	All Other Amusement & Recreation Industries
32300	Rivermere Horse Boarding	Nine Mile Fall	71399050	All Other Amusement & Recreation Industries
33300	Sun Dance Golf Course	Nine Mile Fall	71391002	Golf Courses & Country Clubs
34300	Dry Dock Storage	Valleyford	71393004	Marinas
35300	J R Grennay & Co.	Valleyford	71393003	Marinas
36300	Valley Fitness Inc	Spokane Valley	71394011	Fitness & Recreational Sports Centers
37300	2010 US Figure Skating	Spokane	71399002	All Other Amusement & Recreation Industries
38300	Absolute Fitness	Spokane	71394013	Fitness & Recreational Sports Centers
39300	All About Fitness	Spokane	71394013	Fitness & Recreational Sports Centers
40300	Community Building LLC	Spokane	71394010	Fitness & Recreational Sports Centers
41300	Curves	Spokane	71394011	Fitness & Recreational Sports Centers
42300	Empire Sports Assn	Spokane	71399044	All Other Amusement & Recreation Industries
43300	Laser Quest	Spokane	71399002	All Other Amusement & Recreation Industries
44300	Melanie Morlan-Wellness Cchng	Spokane	71399002	All Other Amusement & Recreation Industries
45300	North Bowl	Spokane	71395001	Bowling Centers
46300	O Z Fitness	Spokane	71394011	Fitness & Recreational Sports Centers
47300	Pioneer Baseball League	Spokane	71399006	All Other Amusement & Recreation Industries
48300	Precision Pilates of Spokane	Spokane	71394023	Fitness & Recreational Sports Centers
49300	Riverfront Park Ice Palace	Spokane	71394025	Fitness & Recreational Sports Centers
50300	Specialty Training	Spokane	71399002	All Other Amusement & Recreation Industries
51300	Spokane Athletic Club	Spokane	71394011	Fitness & Recreational Sports Centers
52300	Spokane Swimming Information	Spokane	71394020	Fitness & Recreational Sports Centers
53300	Wild Walls Climbing Gym & Gear	Spokane	71394009	Fitness & Recreational Sports Centers

APPENDIX

Appendix C cont'd

DesID	COMPANY_NA	PRIMARY_CI	NAICS_CODE	NAICS_DESC
54300	Far East Oriental Spa	Spokane	71394017	Fitness & Recreational Sports Centers
55300	Martin Luther King Jr Family	Spokane	71399014	All Other Amusement & Recreation Industries
56300	Pilates Life	Spokane	71394011	Fitness & Recreational Sports Centers
57300	Spokane Indians	Spokane	71399006	All Other Amusement & Recreation Industries
58300	Virtual Assault Paintball Gms	Spokane	71399031	All Other Amusement & Recreation Industries
59300	Warehouse	Spokane	71394015	Fitness & Recreational Sports Centers
60300	Westside Motorsports	Spokane	71399002	All Other Amusement & Recreation Industries
61300	Challanger Coaching	Spokane	71399002	All Other Amusement & Recreation Industries
62300	Coeur Consulting PLLC	Spokane	71399002	All Other Amusement & Recreation Industries
63300	Snap Fitness	Spokane	71394011	Fitness & Recreational Sports Centers
64300	Curves	Spokane	71394011	Fitness & Recreational Sports Centers
65300	Downriver Municipal Golf Crs	Spokane	71391002	Golf Courses & Country Clubs
66300	Jazzercise Fitness Ctr-Spokane	Spokane	71394011	Fitness & Recreational Sports Centers
67300	Any Season Sports Leagues	Spokane	71394015	Fitness & Recreational Sports Centers
68300	Chester Creek Par 3 Golf Crs	Spokane Valley	71391002	Golf Courses & Country Clubs
69300	Contemporary Fiberglass & Mrn	Spokane Valley	71393003	Marinas
70300	Curves	Spokane Valley	71394011	Fitness & Recreational Sports Centers
71300	Fi's Bodyshop	Spokane Valley	71394013	Fitness & Recreational Sports Centers
72300	Giorgio's Fitness Ctr	Spokane Valley	71394011	Fitness & Recreational Sports Centers
73300	Roller Valley Skate Ctr	Spokane Valley	71394025	Fitness & Recreational Sports Centers
74300	Skyline Inland Northwest	Spokane Valley	71399002	All Other Amusement & Recreation Industries
75300	Spokane Valley Skating & Hcky	Spokane Valley	71394012	Fitness & Recreational Sports Centers
76300	Stroh's Super Sportz	Spokane Valley	71394011	Fitness & Recreational Sports Centers
77300	Accent Fiberglass	Spokane	71393003	Marinas
78300	Digits Relaxation Spa	Spokane	71394017	Fitness & Recreational Sports Centers
79300	Northeast Community Ctr	Spokane	71394010	Fitness & Recreational Sports Centers
80300	Relational Riding Academy	Spokane	71399050	All Other Amusement & Recreation Industries
81300	Spokane Amatuer Softball Assn	Spokane	71399044	All Other Amusement & Recreation Industries

APPENDIX

Appendix C cont'd

DesID	COMPANY_NA	PRIMARY_CI	NAICS_CODE	NAICS_DESC
82300	A-1 Storage	Spokane	71393004	Marinas
83300	Double Eagle Stables Inc	Spokane	71399050	All Other Amusement & Recreation Industries
84300	Eagles Ice-A-Rena	Spokane	71394025	Fitness & Recreational Sports Centers
85300	Global Fitness	Spokane	71394011	Fitness & Recreational Sports Centers
86300	Indian Trail Fitness Ctr	Spokane	71394011	Fitness & Recreational Sports Centers
87300	Lady Raven Stables LLC	Spokane	71399050	All Other Amusement & Recreation Industries
88300	Lilac Lanes	Spokane	71395001	Bowling Centers
89300	North Park Racquet Club	Spokane	71394011	Fitness & Recreational Sports Centers
90300	Northside Fitness	Spokane	71394011	Fitness & Recreational Sports Centers
91300	Phytt Plus Health & Fitness	Spokane	71394011	Fitness & Recreational Sports Centers
92300	Rapid Assault Paintball	Spokane	71399031	All Other Amusement & Recreation Industries
93300	Spokane Americans Youth Hockey	Spokane	71394012	Fitness & Recreational Sports Centers
94300	Trifit Training	Spokane	71394013	Fitness & Recreational Sports Centers
95300	Wandermere Golf Course	Spokane	71391002	Golf Courses & Country Clubs
96300	Athletic Foundations	Spokane	71399002	All Other Amusement & Recreation Industries
97300	Black Clover Boardsports	Spokane	71399044	All Other Amusement & Recreation Industries
98300	Spokane Sports Assn	Spokane	71399044	All Other Amusement & Recreation Industries
99300	Anytime Fitness	Spokane Valley	71394011	Fitness & Recreational Sports Centers
100300	Horizon Fence Co.	Spokane Valley	71399002	All Other Amusement & Recreation Industries
101300	Marine Rigging & Boat Repair	Spokane Valley	71393003	Marinas
102300	Park Road Pool	Spokane Valley	71394020	Fitness & Recreational Sports Centers
103300	Spokane Athletic Club-Valley	Spokane Valley	71394011	Fitness & Recreational Sports Centers
104300	Valley Bowl Inc.	Spokane Valley	71395001	Bowling Centers
105300	Painted Hills Golf Course	Spokane Valley	71391002	Golf Courses & Country Clubs
106300	O Z Fitness	Spokane Valley	71394011	Fitness & Recreational Sports Centers
107300	Powersports & More	Spokane Valley	71399002	All Other Amusement & Recreation Industries
108300	Pure Med Spa	Spokane Valley	71394017	Fitness & Recreational Sports Centers
109300	Winter & Water Sports	Spokane Valley	71399044	All Other Amusement & Recreation Industries

APPENDIX

Appendix C cont'd

DesID	COMPANY_NA	PRIMARY_CI	NAICS_CODE	NAICS_DESC
110300	Brad's Storage Inc.	Spokane	71393004	Marinas
111300	Eastwick Hunters & Jumpers	Spokane	71399050	All Other Amusement & Recreation Industries
112300	Esmeralda Golf Course	Spokane	71391002	Golf Courses & Country Clubs
113300	Fairview Stables	Spokane	71399050	All Other Amusement & Recreation Industries
114300	H & S Certified Marine Svc	Spokane	71393003	Marinas
115300	Hard Rock Motorsports	Spokane	71399002	All Other Amusement & Recreation Industries
116300	Next Step Life Coaching	Spokane	71399002	All Other Amusement & Recreation Industries
117300	Randolph Speed & Marine	Spokane	71393003	Marinas
118300	Above the Line	Spokane	71394013	Fitness & Recreational Sports Centers
119300	Ajuva Medical Spa	Spokane	71394017	Fitness & Recreational Sports Centers
120300	Anytime Fitness	Spokane	71394011	Fitness & Recreational Sports Centers
121300	Curves	Spokane	71394011	Fitness & Recreational Sports Centers
122300	Curves	Spokane	71394011	Fitness & Recreational Sports Centers
123300	Global Fitness for Women	Spokane	71394011	Fitness & Recreational Sports Centers
124300	Gold's Gym	Spokane	71394011	Fitness & Recreational Sports Centers
125300	Jezreel Fitness	Spokane	71394011	Fitness & Recreational Sports Centers
126300	Northwest Winter Sport Fndtn	Spokane	71392002	Skiing Facilities
127300	Oz Fitness	Spokane	71394011	Fitness & Recreational Sports Centers
128300	Pattison's North Skating Rink	Spokane	71394025	Fitness & Recreational Sports Centers
129300	Pine Acres Par 3	Spokane	71391002	Golf Courses & Country Clubs
130300	Spokane Country Club	Spokane	71391001	Golf Courses & Country Clubs
131300	Wonderland Family Fun Ctr	Spokane	71399001	All Other Amusement & Recreation Industries
132300	Spokane Indians Baseball Club	Spokane	71399006	All Other Amusement & Recreation Industries
133300	Big Daddy's Casino	Spokane	71395001	Bowling Centers
134300	Blue Oval Co.	Spokane	71393003	Marinas
135300	Curves	Spokane	71394011	Fitness & Recreational Sports Centers
136300	Fitness Together	Spokane	71394013	Fitness & Recreational Sports Centers
137300	Gold's Gym	Spokane	71394011	Fitness & Recreational Sports Centers

Appendix C cont'd

DesID	COMPANY_NA	PRIMARY_CI	NAICS_CODE	NAICS_DESC
138300	Hangman Valley Golf Course	Spokane	71391002	Golf Courses & Country Clubs
139300	Manito Golf & Country Club	Spokane	71391001	Golf Courses & Country Clubs
140300	Natural Ability	Spokane	71394013	Fitness & Recreational Sports Centers
141300	O Z Fitness	Spokane	71394011	Fitness & Recreational Sports Centers
142300	B L Stables	Spokane	71399050	All Other Amusement & Recreation Industries
143300	Busy Bee Ranch & Equestrian	Spokane	71399050	All Other Amusement & Recreation Industries
144300	Creek at Qualchan Golf Course	Spokane	71391002	Golf Courses & Country Clubs
145300	Eagle Ridge Short Course	Spokane	71391002	Golf Courses & Country Clubs
146300	Elite Fitness	Spokane	71394011	Fitness & Recreational Sports Centers
147300	Indian Canyon Golf Course	Spokane	71391002	Golf Courses & Country Clubs
148300	Medicine Horse Stables	Spokane	71399050	All Other Amusement & Recreation Industries
149300	Prairie Sky Equestrian Ctr	Spokane	71399050	All Other Amusement & Recreation Industries
150300	Spokane Equestrian Ctr	Spokane	71399050	All Other Amusement & Recreation Industries
151300	Spokane Sport Horse Farm	Spokane	71399050	All Other Amusement & Recreation Industries
152300	West Plains Fitness	Spokane	71394011	Fitness & Recreational Sports Centers
153300	West Plains Marine & Rvllc	Spokane	71393003	Marinas
154300	Spokane Pony Baseball	Spokane	71399006	All Other Amusement & Recreation Industries
155300	Spokane Rifle Club	Spokane	71399038	All Other Amusement & Recreation Industries

APPENDIX

Appendix D: List of all Healthcare Facilities within the Study Area

DesID	NAME	City	NAICS_CODE	NAICS_DESC
4400	Rimrock Estates	Chattaroy	62311016	Nursing Care Facilities
5400	Apex Physical Therapy	Cheney	62111108	Offices of Physicians, Except Mental Health
7400	Cheney Assisted Living	Cheney	62311016	Nursing Care Facilities
8400	Cheney Dental Care	Cheney	62121003	Offices of Dentists
11400	Cheney Medical Ctr	Cheney	62111107	Offices of Physicians, Except Mental Health
12400	Cheney Physical Therapy	Cheney	62134007	Offices of Specialty Therapists
13400	Cheney Spinal Care Ctr	Cheney	62149301	Freestanding Emergency Medical Centers
14400	Collins Family Dentistry	Cheney	62121003	Offices of Dentists
15400	Stacy Damiano	Cheney	62134007	Offices of Specialty Therapists
16400	Family Vision Ctr	Cheney	62132003	Offices of Optometrists
17400	First Street Cheney Physical	Cheney	62134007	Offices of Specialty Therapists
19400	Ichoice	Cheney	62141005	Family Planning Centers
20400	Paula Mays	Cheney	62134007	Offices of Specialty Therapists
23400	Greenacres Chiropractic Clinic	Spokane Valley	62131002	Offices of Chiropractors
26400	Spokane Valley Good Samaritan	Spokane Valley	62311016	Nursing Care Facilities
27400	Seniors in Motion	Cheney	62311016	Nursing Care Facilities
28400	Robert I Stockton DDS	Cheney	62121003	Offices of Dentists
32400	Youthful Horizons	Cheney	62134007	Offices of Specialty Therapists
33400	Alpha Omega Sonography	Colbert	62151202	Diagnostic Imaging Centers
37400	Deer Park Nutrition	Deer Park	62139904	Offices of Misc Health Practitioners
41400	Binns Family Chiropractic	Liberty Lake	62131002	Offices of Chiropractors
42400	Casey Family Dental	Liberty Lake	62121003	Offices of Dentists
43400	Clearview Mobile Eye Care	Liberty Lake	62132003	Offices of Optometrists
44400	Condon & Condon	Liberty Lake	62121003	Offices of Dentists
46400	Cullings Family Dentistry	Liberty Lake	62121003	Offices of Dentists

Appendix D cont'd

DesID	NAME	City	NAICS_CODE	NAICS_DESC
47400	Family Chiropractic	Liberty Lake	62131002	Offices of Chiropractors
48400	First Care	Liberty Lake	62149302	Freestanding Emergency Medical Centers
51400	Liberty Lake Dental Care	Liberty Lake	62121003	Offices of Dentists
52400	Liberty Lake Family & Sports	Liberty Lake	62111107	Offices of Physicians, Except Mental Health
53400	Liberty Lake Family Dentistry	Liberty Lake	62121003	Offices of Dentists
54400	Liberty Lake Orthodontics	Liberty Lake	62121003	Offices of Dentists
58400	Neck & Back Decompression Ctr	Liberty Lake	62131002	Offices of Chiropractors
59400	Noreen O' Meara Lac	Liberty Lake	62139901	Offices of Misc Health Practitioners
61400	CGM Consulting Svc	Mead	62151116	Medical Laboratories
64400	Haven Homes	Medical Lake	62311008	Nursing Care Facilities
65400	J A Medical Ctr	Medical Lake	62111107	Offices of Physicians, Except Mental Health
67400	Medical Lake Family Practice	Medical Lake	62111107	Offices of Physicians, Except Mental Health
68400	Medical Lake Physical Therapy	Medical Lake	62134007	Offices of Specialty Therapists
73400	Northwest Office Anesthesia	Spangle	62121003	Offices of Dentists
75400	Spokane Birth & Women's Health	Valleyford	62139956	Offices of Misc Health Practitioners
76400	Back to Basics Chiropractic	Spokane Valley	62131002	Offices of Chiropractors
77400	Chiropractic Wellness Ctr	Spokane Valley	62131002	Offices of Chiropractors
81400	Experience Health Chiropractic	Spokane Valley	62131002	Offices of Chiropractors
82400	Fred Meyer Nutrition	Spokane Valley	62139925	Offices of Misc Health Practitioners
85400	Rare Care Adult Family Home	Veradale	62311016	Nursing Care Facilities
90400	Lighthouse Physical Therapy	Spokane Valley	62134007	Offices of Specialty Therapists
92400	Kathrine Olson DDS	Spokane Valley	62121003	Offices of Dentists
93400	Rodkey Family Dentistry	Spokane Valley	62121003	Offices of Dentists
95400	Spokane Vision	Spokane Valley	62132003	Offices of Optometrists
96400	U S Healthworks	Spokane Valley	62149301	Freestanding Emergency Medical Centers
97400	US Healthworks Physical Thrpy	Spokane Valley	62134007	Offices of Specialty Therapists
98400	Penny C Walpole DDS	Spokane Valley	62121003	Offices of Dentists
99400	Acupuncture Clinic of Spokane	Spokane	62139901	Offices of Misc Health Practitioners

APPENDIX

Appendix D cont'd

DesID	NAME	City	NAICS_CODE	NAICS_DESC
101400	Adoption Legal Svc	Spokane	62139901	Offices of Misc Health Practitioners
102400	Advantage Physical Therapy	Spokane	62134007	Offices of Specialty Therapists
104400	American Medical Response	Spokane	62191002	Ambulance Svcs
109400	Bonnie S Baker PHD	Spokane	62139936	Offices of Misc Health Practitioners
110400	Richard A Bass DDS	Spokane	62121003	Offices of Dentists
112400	Behavioral Assessments	Spokane	62139936	Offices of Misc Health Practitioners
113400	Behavioral Health Northwest	Spokane	62139936	Offices of Misc Health Practitioners
114400	Behaviorial Medicine Svc	Spokane	62139936	Offices of Misc Health Practitioners
115400	Beneficial in-Home Care	Spokane	62161001	Home Health Care Svcs
116400	Berg & Sorensen	Spokane	62111107	Offices of Physicians, Except Mental Health
118400	Loy Mary Blair PHD	Spokane	62139936	Offices of Misc Health Practitioners
119400	Marsha Blasingame MD	Spokane	62111107	Offices of Physicians, Except Mental Health
120400	Brown & Assoc	Spokane	62139936	Offices of Misc Health Practitioners
121400	Home Instead Senior Care	Spokane	62161001	Home Health Care Svcs
126400	O C Olson Diabetes Education	Spokane	62199921	Misc Ambulatory Health Care Svcs
128400	Becky J Oos	Spokane	62139901	Offices of Misc Health Practitioners
130400	Outreach Center	Spokane	62199901	Misc Ambulatory Health Care Svcs
132400	Chesterfield Health Svc Inc	Spokane	62161001	Home Health Care Svcs
133400	Childbirth & Parenting Alone	Spokane	62141005	Family Planning Centers
135400	Chiropractic Clinic	Spokane	62131002	Offices of Chiropractors
136400	Christ Clinic	Spokane	62111107	Offices of Physicians, Except Mental Health
137400	Kenneth H Coleman MD	Spokane	62111107	Offices of Physicians, Except Mental Health
140400	Community Health Assn-Spokane	Spokane	62199921	Misc Ambulatory Health Care Svcs
147400	Dee A Myers MS	Spokane	62139905	Offices of Misc Health Practitioners
152400	Family Medicine	Spokane	62111107	Offices of Physicians, Except Mental Health
154400	Family Service Spokane	Spokane	62221001	Psychiatric & Substance Abuse Hospitals
155400	FHS Healthcare Systems	Spokane	62199921	Misc Ambulatory Health Care Svcs
157400	Fourth Ave Chiropractic	Spokane	62131002	Offices of Chiropractors

APPENDIX

Appendix D cont'd

DesID	NAME	City	NAICS_CODE	NAICS_DESC
158400	Jennifer Gallis	Spokane	62139901	Offices of Misc Health Practitioners
159400	Gen Prime Inc	Spokane	62151109	Medical Laboratories
161400	Christine Guzzardo PHD	Spokane	62139936	Offices of Misc Health Practitioners
162400	Havenwood Caregiver Svc	Spokane	62161001	Home Health Care Svcs
163400	Jeffrey G Hedge MD	Spokane	62111107	Offices of Physicians, Except Mental Health
165400	Inland Northwest Blood Ctr	Spokane	62199101	Blood & Organ Banks
166400	Innovative Wellness Solutions	Spokane	62131002	Offices of Chiropractors
168400	Intrepid USA Healthcare Svc	Spokane	62139920	Offices of Misc Health Practitioners
169400	Joan Chase ARNPCS	Spokane	62221001	Psychiatric & Substance Abuse Hospitals
170400	John R Tiffany Dental Svc	Spokane	62121003	Offices of Dentists
172400	Kristi Ketz PHD	Spokane	62139936	Offices of Misc Health Practitioners
174400	Ladonna Remy Msw Licsw	Spokane	62133003	Offices of Mental Health Practitioners
179400	Longivity Health Ctr	Spokane	62139901	Offices of Misc Health Practitioners
180400	Luchini & Luchini	Spokane	62121003	Offices of Dentists
182400	Magnolia Assessments & Mental	Spokane	62221001	Psychiatric & Substance Abuse Hospitals
185400	Rudyard Mc Kennon DDS	Spokane	62121003	Offices of Dentists
187400	John F Mc Rae PHD	Spokane	62139936	Offices of Misc Health Practitioners
188400	Maxim Healthcare Svc	Spokane	62161001	Home Health Care Svcs
192400	Neuroeducation	Spokane	62139903	Offices of Misc Health Practitioners
193400	Newport Audiology	Spokane	62134001	Offices of Specialty Therapists
195400	NAMI-SPOKANE	Spokane	62221001	Psychiatric & Substance Abuse Hospitals
196400	Native Health of Spokane	Spokane	62149301	Freestanding Emergency Medical Centers
197400	Northwest Behavioral Health	Spokane	62139936	Offices of Misc Health Practitioners
198400	Amy Paris PHD	Spokane	62139936	Offices of Misc Health Practitioners
199400	Parks Medical Corp	Spokane	62111107	Offices of Physicians, Except Mental Health
200400	Parkside Physical Therapy	Spokane	62134007	Offices of Specialty Therapists
201400	Pathology Assoicates	Spokane	62111107	Offices of Physicians, Except Mental Health
203400	PEOPLE'S Clinic	Spokane	62149301	Freestanding Emergency Medical Centers

APPENDIX

Appendix D cont'd

DesID	NAME	City	NAICS_CODE	NAICS_DESC
204400	Peterson & Peterson	Spokane	62121003	Offices of Dentists
205400	Holly H Peterson DDS	Spokane	62121003	Offices of Dentists
207400	Corey L Plaster DDS	Spokane	62121003	Offices of Dentists
208400	Pollack & Assoc	Spokane	62139936	Offices of Misc Health Practitioners
209400	Prosser Dentistry	Spokane	62121003	Offices of Dentists
210400	Psychological Services Spokane	Spokane	62139936	Offices of Misc Health Practitioners
213400	Richard Christy Msw Licsw	Spokane	62133003	Offices of Mental Health Practitioners
214400	Pamela S Ridgway PHD	Spokane	62139936	Offices of Misc Health Practitioners
217400	Volwiler Counseling	Spokane	62139936	Offices of Misc Health Practitioners
219400	Earl L Whittaker DDS	Spokane	62121003	Offices of Dentists
225400	ZLB Plasma Svc	Spokane	62199101	Blood & Organ Banks
226400	A New Hope Inc	Spokane	62139920	Offices of Misc Health Practitioners
230400	Santosha Wellness	Spokane	62199926	Misc Ambulatory Health Care Svcs
231400	Scottish Rite Ctr-Childhood	Spokane	62134009	Offices of Specialty Therapists
233400	SNAP	Spokane	62133003	Offices of Mental Health Practitioners
236400	Spokane Community Care	Spokane	62199921	Misc Ambulatory Health Care Svcs
239400	Spokane Falls Family Clinic	Spokane	62111107	Offices of Physicians, Except Mental Health
240400	Spokane Medical Direct	Spokane	62149301	Freestanding Emergency Medical Centers
242400	Spokane Regional Health Dist	Spokane	62199901	Misc Ambulatory Health Care Svcs
243400	Marc Suffis MD	Spokane	62111107	Offices of Physicians, Except Mental Health
244400	Patrick O Tennican MD	Spokane	62111107	Offices of Physicians, Except Mental Health
246400	American Drug Testing	Spokane	62151103	Medical Laboratories
249400	American Mobile Drug Testing	Spokane	62151103	Medical Laboratories
252400	Asbell Professional Group	Spokane	62139936	Offices of Misc Health Practitioners
254400	Robert G Benedetti MD	Spokane	62111107	Offices of Physicians, Except Mental Health
260400	Cancer Care Northwest PS	Spokane	62111107	Offices of Physicians, Except Mental Health
262400	Abundant Wellness Ctr	Spokane	62199927	Misc Ambulatory Health Care Svcs
264400	John A Adams MD	Spokane	62111107	Offices of Physicians, Except Mental Health

APPENDIX

Appendix D cont'd

DesID	NAME	City	NAICS_CODE	NAICS_DESC
265400	Addus Healthcare	Spokane	62161001	Home Health Care Svcs
268400	Alder Family Chiropractic	Spokane	62131002	Offices of Chiropractors
269400	Ronald J Douglas DPM	Spokane	62111107	Offices of Physicians, Except Mental Health
270400	East Central Primary Care Clnc	Spokane	62149301	Freestanding Emergency Medical Centers
275400	Evergreen Club	Spokane	62221001	Psychiatric & Substance Abuse Hospitals
280400	Champions Sports Medicine	Spokane	62111108	Offices of Physicians, Except Mental Health
282400	Cholesterol Testing Ctr	Spokane	62151106	Medical Laboratories
285400	Brooke M Cloninger DDS	Spokane	62121003	Offices of Dentists
286400	Robert L Cooper MD	Spokane	62111107	Offices of Physicians, Except Mental Health
289400	Day Chiropractic Clinic	Spokane	62131002	Offices of Chiropractors
290400	Deseve & Stevens	Spokane	62139936	Offices of Misc Health Practitioners
292400	T Daniel Dibble MD	Spokane	62111107	Offices of Physicians, Except Mental Health
293400	Division Street Physical Thrpy	Spokane	62134007	Offices of Specialty Therapists
295400	Judith A Heusner MD	Spokane	62111107	Offices of Physicians, Except Mental Health
296400	Illa Hilliard	Spokane	62139923	Offices of Misc Health Practitioners
298400	Inland Eye Ctr	Spokane	62111107	Offices of Physicians, Except Mental Health
299400	Inland Imaging	Spokane	62111107	Offices of Physicians, Except Mental Health
300400	Inland Imaging	Spokane	62111107	Offices of Physicians, Except Mental Health
301400	Inland Medical Evaluation	Spokane	62111107	Offices of Physicians, Except Mental Health
302400	Inland Northwest Renal Care	Spokane	62111107	Offices of Physicians, Except Mental Health
303400	Inland Orthopaedics of Spokane	Spokane	62111107	Offices of Physicians, Except Mental Health
305400	Philip A Lenoue DC	Spokane	62131002	Offices of Chiropractors
308400	Liberty Park Family Dentistry	Spokane	62121003	Offices of Dentists
309400	Life Care Ctr of America	Spokane	62311016	Nursing Care Facilities
310400	Mac Kay Meyer & Hahn	Spokane	62111107	Offices of Physicians, Except Mental Health
312400	Medical Reserve Corps	Spokane	62149301	Freestanding Emergency Medical Centers
313400	Michelle M White PHD & Assoc	Spokane	62139936	Offices of Misc Health Practitioners
314400	Moffitt Children's Dentistry	Spokane	62121003	Offices of Dentists

APPENDIX

Appendix D cont'd

DesID	NAME	City	NAICS_CODE	NAICS_DESC
316400	Neuro Physiology Northwest	Spokane	62139936	Offices of Misc Health Practitioners
317400	Neurology Associates-Spokane	Spokane	62111107	Offices of Physicians, Except Mental Health
318400	Northwest Medical Rehab	Spokane	62111107	Offices of Physicians, Except Mental Health
319400	Occupational Health Solutions	Spokane	62134005	Offices of Specialty Therapists
320400	Occupational Medicine Assoc	Spokane	62111107	Offices of Physicians, Except Mental Health
321400	Alfonso Oliva MD	Spokane	62111107	Offices of Physicians, Except Mental Health
322400	Open MRI Diagnostics	Spokane	62111107	Offices of Physicians, Except Mental Health
323400	Pacific Cataract & Laser Inst	Spokane	62111107	Offices of Physicians, Except Mental Health
325400	Phipps Orthodontics	Spokane	62121003	Offices of Dentists
326400	Pittman Chiropractic Clinic	Spokane	62131002	Offices of Chiropractors
327400	Plastic Surgicentre	Spokane	62111107	Offices of Physicians, Except Mental Health
328400	Powder Basin Accociates Ltd.	Spokane	62111107	Offices of Physicians, Except Mental Health
330400	Providence Health & Svc	Spokane	62199921	Misc Ambulatory Health Care Svcs
331400	Psychlatric Clinical Nurse	Spokane	62139920	Offices of Misc Health Practitioners
332400	John J Rademacher MD	Spokane	62111107	Offices of Physicians, Except Mental Health
334400	Connie S Raybuck PHD	Spokane	62139936	Offices of Misc Health Practitioners
335400	Rehab Associates	Spokane	62111107	Offices of Physicians, Except Mental Health
336400	Retina Associates of Spokane	Spokane	62111107	Offices of Physicians, Except Mental Health
338400	Rockwood Clinic Imaging	Spokane	62111107	Offices of Physicians, Except Mental Health
345400	Spokane District Dental Societ	Spokane	62121003	Offices of Dentists
346400	Spokane Healing Arts	Spokane	62139933	Offices of Misc Health Practitioners
347400	Spokane Mental Health	Spokane	62111107	Offices of Physicians, Except Mental Health
349400	Spokane Mental Health-Child	Spokane	62221001	Psychiatric & Substance Abuse Hospitals
351400	Spokane Veterans Homes	Spokane	62311016	Nursing Care Facilities
352400	St Joseph Care Ctr	Spokane	62311016	Nursing Care Facilities
353400	St Joseph Family Ctr	Spokane	62221001	Psychiatric & Substance Abuse Hospitals
354400	St Luke's Cardiopulmonary	Spokane	62111107	Offices of Physicians, Except Mental Health
355400	St Luke's Rehab Institute	Spokane	62134007	Offices of Specialty Therapists

Appendix D cont'd

DesID	NAME	City	NAICS_CODE	NAICS_DESC
358400	Summit Rehab Assoc	Spokane	62134007	Offices of Specialty Therapists
359400	Jay Toews	Spokane	62139936	Offices of Misc Health Practitioners
363400	VNA Home Health Care Svc	Spokane	62139920	Offices of Misc Health Practitioners
367400	Acupuncture & Naturopathic	Spokane	62149301	Freestanding Emergency Medical Centers
368400	Alijen Consulting LLC	Spokane	62221001	Psychiatric & Substance Abuse Hospitals
371400	Richard H Bale MD	Spokane	62111107	Offices of Physicians, Except Mental Health
373400	Allen D Bostwick PHD	Spokane	62139936	Offices of Misc Health Practitioners
374400	Cannon Hill Dental Clinic	Spokane	62121003	Offices of Dentists
375400	Children's Choice	Spokane	62121003	Offices of Dentists
380400	Marcus De Wood MD	Spokane	62111107	Offices of Physicians, Except Mental Health
381400	Jodi W Funk DDS	Spokane	62121003	Offices of Dentists
383400	Grand Pediatrics	Spokane	62111107	Offices of Physicians, Except Mental Health
384400	Grande Smiles	Spokane	62121003	Offices of Dentists
386400	Charles H Greensword DC	Spokane	62131002	Offices of Chiropractors
387400	Andrew Haffey PHD	Spokane	62139936	Offices of Misc Health Practitioners
390400	Robert I Hustrulid MD	Spokane	62111107	Offices of Physicians, Except Mental Health
391400	Integrative Medicine Assoc	Spokane	62111107	Offices of Physicians, Except Mental Health
392400	Rebecca Kemnitz	Spokane	62139901	Offices of Misc Health Practitioners
393400	Lavander Sage Wellness Ctr	Spokane	62199952	Misc Ambulatory Health Care Svcs
395400	Manito Family Dentistry	Spokane	62121003	Offices of Dentists
398400	Sean Mee PHD	Spokane	62111107	Offices of Physicians, Except Mental Health
399400	Milestones	Spokane	62134009	Offices of Specialty Therapists
400400	Stephen Mills DDS	Spokane	62121003	Offices of Dentists
403400	Physical Therapy Assoc	Spokane	62134007	Offices of Specialty Therapists
404400	Porter Chiropractic Ctr	Spokane	62131002	Offices of Chiropractors
406400	Quiesco Anesthesia Svc	Spokane	62111101	Offices of Physicians, Except Mental Health
408400	Charles L Regalado DDS	Spokane	62121003	Offices of Dentists
409400	Restore Vision Ctr	Spokane	62111107	Offices of Physicians, Except Mental Health

APPENDIX

Appendix D cont'd

DesID	NAME	City	NAICS_CODE	NAICS_DESC
410400	Stanley A Sargent DDS	Spokane	62121003	Offices of Dentists
411400	Jay H Sciuchetti DDS	Spokane	62121003	Offices of Dentists
412400	Slack & Combs Orthodontics	Spokane	62121003	Offices of Dentists
413400	South Hill Family Dentistry	Spokane	62121003	Offices of Dentists
414400	South Hill Foot & Ankle Clinic	Spokane	62139103	Offices of Podiatrists
415400	South Hill Physical Therapy	Spokane	62134007	Offices of Specialty Therapists
416400	South Hill Spinal Care	Spokane	62131002	Offices of Chiropractors
419400	Spokane Homeopathic Clinic	Spokane	62139933	Offices of Misc Health Practitioners
420400	Tailwind Physical Therapy	Spokane	62134007	Offices of Specialty Therapists
422400	U S Healthworks Medical Clinic	Spokane	62149301	Freestanding Emergency Medical Centers
423400	Waterford Home Care	Spokane	62161001	Home Health Care Svcs
424400	Waterford On South Hill	Spokane	62311016	Nursing Care Facilities
426400	A Center for Health & Wellness	Spokane	62111107	Offices of Physicians, Except Mental Health
427400	A Turning Pointe Physical Thry	Spokane	62134007	Offices of Specialty Therapists
428400	Advanced Surgical Othopedics	Spokane	62111107	Offices of Physicians, Except Mental Health
429400	Aesthetic Plastic Surgical Ctr	Spokane	62111107	Offices of Physicians, Except Mental Health
431400	American College-Prsthdntsts	Spokane	62121003	Offices of Dentists
432400	Jonathan M Anderson MD	Spokane	62111107	Offices of Physicians, Except Mental Health
434400	Cristian Andronic MD	Spokane	62111107	Offices of Physicians, Except Mental Health
436400	Arthritis Northwest	Spokane	62111107	Offices of Physicians, Except Mental Health
437400	Aspen Oral & Facial	Spokane	62121003	Offices of Dentists
438400	Associated Surgeons	Spokane	62111107	Offices of Physicians, Except Mental Health
439400	George W Bagby MD	Spokane	62111107	Offices of Physicians, Except Mental Health
441400	Breast Evaluation Ctr	Spokane	62111107	Offices of Physicians, Except Mental Health
442400	James D Breeden DDS	Spokane	62121003	Offices of Dentists
445400	Carl F Brunjes MD	Spokane	62111107	Offices of Physicians, Except Mental Health
446400	Cancer Care Ctr	Spokane	62111107	Offices of Physicians, Except Mental Health
447400	Cancer Care Northwest	Spokane	62111107	Offices of Physicians, Except Mental Health

Appendix D cont'd

DesID	NAME	City	NAICS_CODE	NAICS_DESC
448400	Cardiovascular Services	Spokane	62111107	Offices of Physicians, Except Mental Health
451400	Center for Maternal Fetal Mdcn	Spokane	62111107	Offices of Physicians, Except Mental Health
452400	Center for Reproductive Endo	Spokane	62111107	Offices of Physicians, Except Mental Health
454400	Child Neurology	Spokane	62111107	Offices of Physicians, Except Mental Health
456400	Collins Oral & Maxillofacial	Spokane	62121003	Offices of Dentists
459400	Constance Copetas DDS	Spokane	62121003	Offices of Dentists
460400	Cornerstone Psychologists	Spokane	62139936	Offices of Misc Health Practitioners
461400	Daniel R Coulston MD	Spokane	62111107	Offices of Physicians, Except Mental Health
463400	Daniel Phillips & Assoc	Spokane	62111107	Offices of Physicians, Except Mental Health
464400	David A Scott Inc.	Spokane	62139936	Offices of Misc Health Practitioners
465400	Deaconess Behavioral Medicine	Spokane	62111107	Offices of Physicians, Except Mental Health
466400	Deaconess Breat Evaluation Ctr	Spokane	62151203	Diagnostic Imaging Centers
467400	Deaconess Hospital-Emergency	Spokane	62111107	Offices of Physicians, Except Mental Health
468400	Gordon W Decker MD	Spokane	62111107	Offices of Physicians, Except Mental Health
469400	Samuel W Delaney Jr PHD	Spokane	62139936	Offices of Misc Health Practitioners
470400	Lawrence Eastburn MD	Spokane	62111107	Offices of Physicians, Except Mental Health
471400	Endocrine Assoc of Spokane	Spokane	62111107	Offices of Physicians, Except Mental Health
472400	David L Erb PHD	Spokane	62139936	Offices of Misc Health Practitioners
473400	Gilbert R Escandon MD	Spokane	62111107	Offices of Physicians, Except Mental Health
474400	John P Everett MD	Spokane	62111107	Offices of Physicians, Except Mental Health
475400	Family Health Ctr	Spokane	62111107	Offices of Physicians, Except Mental Health
477400	Roger D Fincher MD	Spokane	62111107	Offices of Physicians, Except Mental Health
478400	John F Floyd MD	Spokane	62111107	Offices of Physicians, Except Mental Health
479400	Full Circle Medical Clinic	Spokane	62149301	Freestanding Emergency Medical Centers
480400	Sean W Garman MD	Spokane	62111107	Offices of Physicians, Except Mental Health
481400	Gerald L Gates MD	Spokane	62111107	Offices of Physicians, Except Mental Health
482400	Edward P Gould MD	Spokane	62111107	Offices of Physicians, Except Mental Health
485400	Steven E Gregg DDS	Spokane	62121003	Offices of Dentists

APPENDIX

Appendix D cont'd

DesID	NAME	City	NAICS_CODE	NAICS_DESC
486400	Jeffrey E Hartman MD	Spokane	62111107	Offices of Physicians, Except Mental Health
488400	Heart Attack Prevention Clinic	Spokane	62111107	Offices of Physicians, Except Mental Health
489400	Heart Clinic NW	Spokane	62111107	Offices of Physicians, Except Mental Health
490400	Heart Clinics Northwest	Spokane	62111107	Offices of Physicians, Except Mental Health
491400	Jean F Herzog MD	Spokane	62111107	Offices of Physicians, Except Mental Health
492400	Homeless Resource Ctr	Spokane	62221001	Psychiatric & Substance Abuse Hospitals
494400	House & Mengert	Spokane	62121003	Offices of Dentists
495400	Icard PLLC	Spokane	62139936	Offices of Misc Health Practitioners
496400	Infant Developemental Clinic	Spokane	62111107	Offices of Physicians, Except Mental Health
497400	Inland Cardiology Assoc	Spokane	62111107	Offices of Physicians, Except Mental Health
499400	Inland Empire Gastroenterology	Spokane	62111107	Offices of Physicians, Except Mental Health
500400	Inland Imaging at Sacred Heart	Spokane	62111107	Offices of Physicians, Except Mental Health
501400	Inland Imaging Inc.	Spokane	62151106	Medical Laboratories
502400	Inland Northwest Pediatric	Spokane	62111107	Offices of Physicians, Except Mental Health
503400	Inland NW Regional Perinatal	Spokane	62111107	Offices of Physicians, Except Mental Health
504400	Inland Surgical Assoc Pllc	Spokane	62149810	All Other Outpatient Care Centers
505400	Inland Vascular Institute	Spokane	62111107	Offices of Physicians, Except Mental Health
506400	Johnson Dental	Spokane	62121003	Offices of Dentists
507400	Kidney Care Spokane	Spokane	62111107	Offices of Physicians, Except Mental Health
508400	Kim Krull Acupuncturist	Spokane	62139901	Offices of Misc Health Practitioners
509400	Lifecenter Northwest Donor	Spokane	62199102	Blood & Organ Banks
510400	Jonathan W Lueders MD	Spokane	62111107	Offices of Physicians, Except Mental Health
511400	Marycliff Allergy Specialists	Spokane	62111107	Offices of Physicians, Except Mental Health
513400	Phyllis Mast PHD	Spokane	62139936	Offices of Misc Health Practitioners
515400	Mark Mays PHD	Spokane	62139936	Offices of Misc Health Practitioners
516400	Michael Mc Carthy MD	Spokane	62111107	Offices of Physicians, Except Mental Health
517400	John B Mc Connaughey DDS	Spokane	62121003	Offices of Dentists
518400	Medical Oncology Assoc	Spokane	62111107	Offices of Physicians, Except Mental Health

Appendix D cont'd

DesID	NAME	City	NAICS_CODE	NAICS_DESC
519400	James R Mengert DDS	Spokane	62121003	Offices of Dentists
520400	Todd Merchen MD	Spokane	62111107	Offices of Physicians, Except Mental Health
521400	Mary F Miller PHD	Spokane	62139936	Offices of Misc Health Practitioners
523400	Molina Healthcare of Wa	Spokane	62199921	Misc Ambulatory Health Care Svcs
524400	David G Morgan MD	Spokane	62111107	Offices of Physicians, Except Mental Health
525400	N W Andrology & Cryobank	Spokane	62199950	Misc Ambulatory Health Care Svcs
526400	Neonatology Associates Spokane	Spokane	62111107	Offices of Physicians, Except Mental Health
527400	Neurosurgery Associates	Spokane	62111107	Offices of Physicians, Except Mental Health
528400	NHR Haelan Med Evaluation	Spokane	62199928	Misc Ambulatory Health Care Svcs
529400	NKS Rehab & Wellness Ctr	Spokane	62134007	Offices of Specialty Therapists
530400	Northside Psychiatric Group	Spokane	62111107	Offices of Physicians, Except Mental Health
532400	Northwest Cardio Thoracic	Spokane	62111107	Offices of Physicians, Except Mental Health
533400	Northwest Center-Congenital	Spokane	62111107	Offices of Physicians, Except Mental Health
534400	Northwest Heart & Lung	Spokane	62111107	Offices of Physicians, Except Mental Health
535400	Northwest Heart & Lung	Spokane	62111107	Offices of Physicians, Except Mental Health
536400	Northwest Neurological	Spokane	62111107	Offices of Physicians, Except Mental Health
537400	Northwest Ob-Gyn	Spokane	62111107	Offices of Physicians, Except Mental Health
538400	Northwest Orthopaedic Spec	Spokane	62111107	Offices of Physicians, Except Mental Health
539400	Northwest Pediatric Ophthlmlgy	Spokane	62111107	Offices of Physicians, Except Mental Health
540400	Northwest Renal Svc	Spokane	62111107	Offices of Physicians, Except Mental Health
541400	Ob Gyn Assoc of Spokane	Spokane	62111107	Offices of Physicians, Except Mental Health
542400	Orthopedic & Spine Surgery	Spokane	62111107	Offices of Physicians, Except Mental Health
543400	Jack L Ossello DDS	Spokane	62121003	Offices of Dentists
544400	Deanette L Palmer PHD	Spokane	62139936	Offices of Misc Health Practitioners
545400	Partners With Families & Child	Spokane	62149301	Freestanding Emergency Medical Centers
546400	Pathology Associates Medical	Spokane	62151106	Medical Laboratories
547400	Pediatric Associates-Spokane	Spokane	62111107	Offices of Physicians, Except Mental Health
548400	Pediatric Endocrinology	Spokane	62111107	Offices of Physicians, Except Mental Health

APPENDIX

Appendix D cont'd

DesID	NAME	City	NAICS_CODE	NAICS_DESC
549400	Pediatric Multispecialty Clnc	Spokane	62111107	Offices of Physicians, Except Mental Health
550400	Pediatric Specialities	Spokane	62111107	Offices of Physicians, Except Mental Health
551400	Pediatrix Medical Group	Spokane	62111107	Offices of Physicians, Except Mental Health
552400	Elizabeth Peterson MD	Spokane	62111107	Offices of Physicians, Except Mental Health
553400	Phototherapy & Patch Test Ctr	Spokane	62111107	Offices of Physicians, Except Mental Health
554400	Physician Anesthesia Group	Spokane	62111107	Offices of Physicians, Except Mental Health
555400	Physicians Clinic	Spokane	62111107	Offices of Physicians, Except Mental Health
556400	Physicians Clinic of Spokane	Spokane	62111107	Offices of Physicians, Except Mental Health
557400	Physicians Clinic of Spokane	Spokane	62111107	Offices of Physicians, Except Mental Health
559400	Providence Cancer Ctr	Spokane	62111107	Offices of Physicians, Except Mental Health
560400	Providence Neuroscience Ctr	Spokane	62111107	Offices of Physicians, Except Mental Health
561400	Bridget M Raleigh	Spokane	62139923	Offices of Misc Health Practitioners
562400	Rehabilitation Medicine	Spokane	62111107	Offices of Physicians, Except Mental Health
564400	Donald D Roberts PHD	Spokane	62139936	Offices of Misc Health Practitioners
567400	James A Roubos	Spokane	62139936	Offices of Misc Health Practitioners
568400	Sacred Heart	Spokane	62149301	Freestanding Emergency Medical Centers
569400	Sacred Heart Children's Hosp	Spokane	62211002	General Medical & Surgical Hospitals
570400	Sacred Heart Wound Care Svc	Spokane	62211002	General Medical & Surgical Hospitals
571400	W M Shanks MD	Spokane	62111107	Offices of Physicians, Except Mental Health
572400	Gary Shellerud DDS	Spokane	62121003	Offices of Dentists
573400	Robert Sigman MD	Spokane	62111107	Offices of Physicians, Except Mental Health
575400	Steven V Silverstein MD	Spokane	62111107	Offices of Physicians, Except Mental Health
577400	Spakane Radiation Oncology	Spokane	62139936	Offices of Misc Health Practitioners
579400	Spokane Aids Network	Spokane	62199901	Misc Ambulatory Health Care Svcs
580400	Spokane Allergy & Asthma Clnc	Spokane	62111107	Offices of Physicians, Except Mental Health
581400	Spokane Cardiology	Spokane	62111107	Offices of Physicians, Except Mental Health
582400	Spokane Clinic for Rectal	Spokane	62111107	Offices of Physicians, Except Mental Health
583400	Spokane Dermatology	Spokane	62111107	Offices of Physicians, Except Mental Health

Appendix D cont'd

DesID	NAME	City	NAICS_CODE	NAICS_DESC
584400	Spokane Digestive Disease Ctr	Spokane	62111107	Offices of Physicians, Except Mental Health
585400	Spokane Digestive Disease Ctr	Spokane	62111107	Offices of Physicians, Except Mental Health
586400	Spokane Ear Nose & Throat	Spokane	62111107	Offices of Physicians, Except Mental Health
587400	Spokane Eye Surgical Ctr	Spokane	62111107	Offices of Physicians, Except Mental Health
588400	Spokane Foot Clinic	Spokane	62139103	Offices of Podiatrists
589400	Spokane Headache Clinic	Spokane	62111107	Offices of Physicians, Except Mental Health
590400	Spokane Obstetrics & Gyn	Spokane	62111107	Offices of Physicians, Except Mental Health
591400	Spokane Occupational Therapy	Spokane	62134006	Offices of Specialty Therapists
592400	Spokane Panel Evaluations	Spokane	62111107	Offices of Physicians, Except Mental Health
593400	Spokane Pediatrics Specialist	Spokane	62111107	Offices of Physicians, Except Mental Health
594400	Spokane Psychiatric Clinic	Spokane	62111107	Offices of Physicians, Except Mental Health
595400	Spokane Psychiatry & Psychclgy	Spokane	62111107	Offices of Physicians, Except Mental Health
596400	Spokane Psychology & Neuro	Spokane	62139936	Offices of Misc Health Practitioners
597400	Spokane Radiation Oncology	Spokane	62111107	Offices of Physicians, Except Mental Health
598400	Spokane Radiology Consultants	Spokane	62111107	Offices of Physicians, Except Mental Health
599400	Spokane Respiratory Conslnts	Spokane	62111107	Offices of Physicians, Except Mental Health
600400	Spokane Urology	Spokane	62111107	Offices of Physicians, Except Mental Health
602400	Star Sports Therapy & Rehab	Spokane	62134007	Offices of Specialty Therapists
605400	Surgical Specalists Spokane	Spokane	62111107	Offices of Physicians, Except Mental Health
606400	Surgical Specialists	Spokane	62111107	Offices of Physicians, Except Mental Health
607400	Judy Swanson MD	Spokane	62111107	Offices of Physicians, Except Mental Health
609400	Sharon Underwood PHD	Spokane	62139936	Offices of Misc Health Practitioners
610400	Vedbrat S Vaid MD	Spokane	62111107	Offices of Physicians, Except Mental Health
611400	Washington Medical Memory Clnc	Spokane	62139936	Offices of Misc Health Practitioners
612400	Washington Outpatient Rehab	Spokane	62134006	Offices of Specialty Therapists
613400	Paul M Wert PHD	Spokane	62139936	Offices of Misc Health Practitioners
614400	West Plains Medical	Spokane	62111107	Offices of Physicians, Except Mental Health
615400	Womans Care	Spokane	62111107	Offices of Physicians, Except Mental Health

APPENDIX

Appendix D cont'd

DesID	NAME	City	NAICS_CODE	NAICS_DESC
616400	Women's Health Svc Coach	Spokane	62151203	Diagnostic Imaging Centers
617400	Womens Center Helpline	Spokane	62111107	Offices of Physicians, Except Mental Health
618400	Hershel Zellman MD	Spokane	62111107	Offices of Physicians, Except Mental Health
619400	Zugec Medical Clinic	Spokane	62111107	Offices of Physicians, Except Mental Health
620400	Ability Physical Therapy	Spokane	62134007	Offices of Specialty Therapists
622400	Advanced Chiropractic Clinic	Spokane	62131002	Offices of Chiropractors
623400	Traci Anderson OD	Spokane	62132003	Offices of Optometrists
625400	Anyan Family Dentistry	Spokane	62121003	Offices of Dentists
626400	Ash & Rowan Family Dentistry	Spokane	62121003	Offices of Dentists
627400	Audubon Park Chiropractic	Spokane	62131002	Offices of Chiropractors
628400	Beneficial in Home Care	Spokane	62161001	Home Health Care Svcs
629400	Bond Chiropractic Health Ctr	Spokane	62131002	Offices of Chiropractors
631400	William L Brown MD	Spokane	62111107	Offices of Physicians, Except Mental Health
634400	Victor V Carnell DDS	Spokane	62121003	Offices of Dentists
635400	Edward L Charbonneau DDS	Spokane	62121003	Offices of Dentists
636400	CHAS Clinic	Spokane	62199921	Misc Ambulatory Health Care Svcs
638400	Chiropractic Northside	Spokane	62131002	Offices of Chiropractors
640400	Mary Correll DDS	Spokane	62121003	Offices of Dentists
642400	Deaconess Women's Clinic	Spokane	62139916	Offices of Misc Health Practitioners
645400	Five Mile Family Dentistry	Spokane	62121003	Offices of Dentists
646400	Four Seasons Therapy LLP	Spokane	62134007	Offices of Specialty Therapists
647400	Garland Physical Therapy	Spokane	62134007	Offices of Specialty Therapists
648400	Gentle Touch Chiropractic	Spokane	62131002	Offices of Chiropractors
649400	Robert Granly DDS	Spokane	62121003	Offices of Dentists
650400	Guthrie Chiropractic	Spokane	62131002	Offices of Chiropractors
651400	William E Hallinan DDS	Spokane	62121003	Offices of Dentists
655400	Holistic Dental Ctr	Spokane	62121003	Offices of Dentists
656400	Holistic Family Dentistry	Spokane	62121003	Offices of Dentists

Appendix D cont'd

DesID	NAME	City	NAICS_CODE	NAICS_DESC
659400	Judy's Best Care	Spokane	62311016	Nursing Care Facilities
660400	Jukich Chiropractic Clinic	Spokane	62131002	Offices of Chiropractors
663400	Michael E Kondo DDS	Spokane	62121003	Offices of Dentists
665400	Life Services/Crisis Pregnancy	Spokane	62149301	Freestanding Emergency Medical Centers
666400	Scott M Lindquist DC	Spokane	62131002	Offices of Chiropractors
669400	Manor Care Health Svc	Spokane	62311016	Nursing Care Facilities
670400	Don G Marshall DDS	Spokane	62121003	Offices of Dentists
671400	Martin Family Dental Care	Spokane	62121003	Offices of Dentists
672400	Medical Foot Ctr	Spokane	62111107	Offices of Physicians, Except Mental Health
673400	North Central Care Ctr	Spokane	62311016	Nursing Care Facilities
674400	Northside Family Dentistry	Spokane	62121003	Offices of Dentists
675400	Northside Podiatry	Spokane	62111107	Offices of Physicians, Except Mental Health
676400	Northside Vision Ctr	Spokane	62132003	Offices of Optometrists
677400	Northwest Spine & Posture Ctr	Spokane	62131002	Offices of Chiropractors
679400	Old Firehouse Dental Office	Spokane	62121003	Offices of Dentists
680400	Oral Surgery Plus	Spokane	62121003	Offices of Dentists
682400	Petersen Family Dentistry	Spokane	62121003	Offices of Dentists
684400	Premier Dental-Spokane	Spokane	62121003	Offices of Dentists
685400	Michael Readel DDS	Spokane	62121003	Offices of Dentists
686400	Alyson G Roby MD	Spokane	62111107	Offices of Physicians, Except Mental Health
688400	Rowse's Office	Spokane	62139933	Offices of Misc Health Practitioners
690400	Evan D Schafer DDS	Spokane	62121003	Offices of Dentists
691400	Scott Chiropractic Ctr	Spokane	62131002	Offices of Chiropractors
692400	Sesso Chiropractic	Spokane	62131002	Offices of Chiropractors
693400	Shared Transitions LLC	Spokane	62199927	Misc Ambulatory Health Care Svcs
694400	Shoemaker Chiropractic	Spokane	62131002	Offices of Chiropractors
695400	Sicilia Chiropractic	Spokane	62131002	Offices of Chiropractors
696400	Siloam Acupuncture Clinic	Spokane	62139901	Offices of Misc Health Practitioners

APPENDIX

Appendix D cont'd

DesID	NAME	City	NAICS_CODE	NAICS_DESC
697400	Craig B Simmons DDS	Spokane	62121003	Offices of Dentists
699400	Spinalaid Centers of America	Spokane	62131002	Offices of Chiropractors
700400	Spokane Family Dental	Spokane	62121003	Offices of Dentists
701400	Spokane Foot Clinic	Spokane	62111107	Offices of Physicians, Except Mental Health
703400	Star Medical Inc	Spokane	62111107	Offices of Physicians, Except Mental Health
704400	Thompson Chiropractic Offices	Spokane	62131002	Offices of Chiropractors
707400	US Veterans Medical Ctr	Spokane	62211002	General Medical & Surgical Hospitals
709400	Frank F Vedelago DDS	Spokane	62121003	Offices of Dentists
710400	Windrose Naturopathic Clinic	Spokane	62139933	Offices of Misc Health Practitioners
712400	A Back & Neck Pain Ctr	Spokane Valley	62131002	Offices of Chiropractors
713400	Adult Alternative Accomodation	Spokane Valley	62311016	Nursing Care Facilities
714400	All Valley Medical	Spokane Valley	62111107	Offices of Physicians, Except Mental Health
715400	Alpine Lodge Adult Family Home	Spokane Valley	62311008	Nursing Care Facilities
716400	American College-Prsthdntsts	Spokane Valley	62121003	Offices of Dentists
717400	Associated Family Physicians	Spokane Valley	62111107	Offices of Physicians, Except Mental Health
718400	At Home Care	Spokane Valley	62161001	Home Health Care Svcs
719400	Avenue Dental Care	Spokane Valley	62121003	Offices of Dentists
720400	John T Belknap DDS	Spokane Valley	62121003	Offices of Dentists
721400	Steven E Belknap DDS	Spokane Valley	62121003	Offices of Dentists
722400	Berube Senior Insurance	Spokane Valley	62199921	Misc Ambulatory Health Care Svcs
724400	Barbara Briscoe	Spokane Valley	62133001	Offices of Mental Health Practitioners
725400	John R Burke DDS	Spokane Valley	62121003	Offices of Dentists
729400	Jennifer L Chamberlain	Spokane Valley	62139923	Offices of Misc Health Practitioners
730400	Chas Clinic	Spokane Valley	62199921	Misc Ambulatory Health Care Svcs
732400	Chiropractic & Therapy Assoc	Spokane Valley	62131002	Offices of Chiropractors
735400	Jay R Clark DDS	Spokane Valley	62121003	Offices of Dentists
736400	Colonial Court Asst Living	Spokane Valley	62311016	Nursing Care Facilities
737400	Scott J Crews DDS	Spokane Valley	62121003	Offices of Dentists

Appendix D cont'd

DesID	NAME	City	NAICS_CODE	NAICS_DESC
738400	Crisis Pregnancy Ctr	Spokane Valley	62141005	Family Planning Centers
739400	Dynamic Medical Systems	Spokane Valley	62149301	Freestanding Emergency Medical Centers
740400	Ellingsen/Paxton Orthodontics	Spokane Valley	62121003	Offices of Dentists
741400	Examone	Spokane Valley	62199930	Misc Ambulatory Health Care Svcs
743400	Family Foot Ctr	Spokane Valley	62139103	Offices of Podiatrists
744400	Family Home Care Private Duty	Spokane Valley	62139920	Offices of Misc Health Practitioners
746400	Foot & Ankle Clinic of Spokane	Spokane Valley	62139103	Offices of Podiatrists
747400	Gardens On University	Spokane Valley	62134007	Offices of Specialty Therapists
749400	David C Hagelin DDS	Spokane Valley	62121003	Offices of Dentists
752400	Burton Hart DO	Spokane Valley	62111107	Offices of Physicians, Except Mental Health
753400	Gordon E Hawk DDS	Spokane Valley	62121003	Offices of Dentists
754400	Health First Chiropractic	Spokane Valley	62131002	Offices of Chiropractors
755400	Layne D Hinckley DDS	Spokane Valley	62121003	Offices of Dentists
756400	Allan E Hinkle DDS	Spokane Valley	62121003	Offices of Dentists
758400	Carrie Holliday	Spokane Valley	62139923	Offices of Misc Health Practitioners
760400	Richard W Illes MD	Spokane Valley	62111107	Offices of Physicians, Except Mental Health
761400	In Step Foot & Ankle	Spokane Valley	62111107	Offices of Physicians, Except Mental Health
762400	Independent Services Corp	Spokane Valley	62161001	Home Health Care Svcs
763400	Inland Neurology Pllc	Spokane Valley	62111107	Offices of Physicians, Except Mental Health
764400	Inland Physical Therapy	Spokane Valley	62134007	Offices of Specialty Therapists
765400	Integrated Health Professional	Spokane Valley	62139920	Offices of Misc Health Practitioners
767400	Gary D Keller DDS	Spokane Valley	62121003	Offices of Dentists
770400	Lifecare Solutions	Spokane Valley	62161001	Home Health Care Svcs
772400	Luna Eye Ctr	Spokane Valley	62139959	Offices of Misc Health Practitioners
773400	Douglas S Mac Kay DDS	Spokane Valley	62121003	Offices of Dentists
774400	Mc Donald & Melkers	Spokane Valley	62121003	Offices of Dentists
775400	Medical Foot Ctr	Spokane Valley	62111107	Offices of Physicians, Except Mental Health
777400	Meridian Dental Study Group	Spokane Valley	62121003	Offices of Dentists

APPENDIX

Appendix D cont'd

DesID	NAME	City	NAICS_CODE	NAICS_DESC
778400	Frank A Morgan DDS	Spokane Valley	62121003	Offices of Dentists
779400	N W Eyecare & Laser Ctr	Spokane Valley	62111107	Offices of Physicians, Except Mental Health
780400	Noble Care Adult Family Home	Spokane Valley	62311016	Nursing Care Facilities
781400	North Pines Chiropractic	Spokane Valley	62131002	Offices of Chiropractors
782400	North Pines Counseling	Spokane Valley	62139936	Offices of Misc Health Practitioners
783400	North Pines Dental Care	Spokane Valley	62121003	Offices of Dentists
785400	Pearson & Weary Pain Relief	Spokane Valley	62149301	Freestanding Emergency Medical Centers
786400	Pediatric Therapy Specialists	Spokane Valley	62134006	Offices of Specialty Therapists
787400	Pines Family Medicine	Spokane Valley	62149301	Freestanding Emergency Medical Centers
789400	Point of Origin	Spokane Valley	62139901	Offices of Misc Health Practitioners
792400	Progressive Physical Therapy	Spokane Valley	62134007	Offices of Specialty Therapists
793400	Rapid Clinics	Spokane Valley	62149301	Freestanding Emergency Medical Centers
796400	John B Ryan DDS	Spokane Valley	62121003	Offices of Dentists
800400	Soins De Beaute'	Spokane Valley	62199921	Misc Ambulatory Health Care Svcs
801400	South Pines Chiropractic Ofc	Spokane Valley	62131002	Offices of Chiropractors
802400	Spokane Foot Clinic	Spokane Valley	62139103	Offices of Podiatrists
803400	Spokane Mental Health	Spokane Valley	62221001	Psychiatric & Substance Abuse Hospitals
805400	Spokane Occupational Therapy	Spokane Valley	62134006	Offices of Specialty Therapists
806400	Spokane Oral & Maxillofacial	Spokane Valley	62121003	Offices of Dentists
808400	Spokane Valley Chiropractic	Spokane Valley	62131002	Offices of Chiropractors
809400	Spokane Valley Dental	Spokane Valley	62121003	Offices of Dentists
810400	Spokane Valley Dermatology	Spokane Valley	62111107	Offices of Physicians, Except Mental Health
813400	Sunshine Gardens	Spokane Valley	62311016	Nursing Care Facilities
816400	Michael A Trantow DDS	Spokane Valley	62121003	Offices of Dentists
817400	University Chiropractic Clinic	Spokane Valley	62131002	Offices of Chiropractors
818400	Valley Dialysis Ctr	Spokane Valley	62149301	Freestanding Emergency Medical Centers
820400	Valley Physical Therapy	Spokane Valley	62134007	Offices of Specialty Therapists
822400	Velis & Assoc	Spokane Valley	62121003	Offices of Dentists

Appendix D cont'd

DesID	NAME	City	NAICS_CODE	NAICS_DESC
825400	Robert G Wendel DDS	Spokane Valley	62121003	Offices of Dentists
826400	Stephen O Woodard DDS	Spokane Valley	62121003	Offices of Dentists
827400	ZLB Plasma Svc	Spokane Valley	62199101	Blood & Organ Banks
828400	Advancing Hope Hypnotherapy	Spokane	62139912	Offices of Misc Health Practitioners
830400	Anchor of Light	Spokane	62139938	Offices of Misc Health Practitioners
831400	Associated Dentists	Spokane	62121003	Offices of Dentists
832400	Beacon Hill Dental	Spokane	62121003	Offices of Dentists
834400	Birthright	Spokane	62139916	Offices of Misc Health Practitioners
835400	James C Bonvallet MD	Spokane	62111107	Offices of Physicians, Except Mental Health
837400	C William Britt Jr MD	Spokane	62111107	Offices of Physicians, Except Mental Health
839400	Chiropractic Center	Spokane	62131002	Offices of Chiropractors
840400	Comfort Keepers	Spokane	62161001	Home Health Care Svcs
842400	Damon & Damon	Spokane	62121003	Offices of Dentists
843400	Dermatology Associates	Spokane	62111107	Offices of Physicians, Except Mental Health
844400	Doctor's Clinic	Spokane	62111107	Offices of Physicians, Except Mental Health
846400	Family Dental Care	Spokane	62121003	Offices of Dentists
849400	Glass Chiropractic Clinic	Spokane	62131002	Offices of Chiropractors
851400	Linda Higley PHD	Spokane	62139936	Offices of Misc Health Practitioners
853400	Holy Family Foot & Gait Clinic	Spokane	62111107	Offices of Physicians, Except Mental Health
854400	V Patrick Hughes MD	Spokane	62111107	Offices of Physicians, Except Mental Health
855400	Norman J James MD	Spokane	62111107	Offices of Physicians, Except Mental Health
856400	Amy Johnson PHD	Spokane	62139936	Offices of Misc Health Practitioners
857400	Jorgensen Dietzen # Assoc	Spokane	62139936	Offices of Misc Health Practitioners
861400	Lee & Lee	Spokane	62111107	Offices of Physicians, Except Mental Health
862400	Leoff Health & Welfare Trust	Spokane	62199921	Misc Ambulatory Health Care Svcs
865400	Logan Hurst Health Care	Spokane	62311016	Nursing Care Facilities
868400	Morscheck & Morscheck	Spokane	62139936	Offices of Misc Health Practitioners
869400	Keith A Morton MD	Spokane	62111107	Offices of Physicians, Except Mental Health

APPENDIX

Appendix D cont'd

DesID	NAME	City	NAICS_CODE	NAICS_DESC
871400	North Spokane Pulmonary Clinic	Spokane	62111107	Offices of Physicians, Except Mental Health
872400	North Spokane Surgeons	Spokane	62111107	Offices of Physicians, Except Mental Health
873400	North Spokane Women's Clinic	Spokane	62111107	Offices of Physicians, Except Mental Health
874400	Northeast CHAS Clinic	Spokane	62199921	Misc Ambulatory Health Care Svcs
875400	Northside Family Medicine	Spokane	62111107	Offices of Physicians, Except Mental Health
876400	Northside Family Physicians	Spokane	62111107	Offices of Physicians, Except Mental Health
877400	Northtown Vision Clinic	Spokane	62132003	Offices of Optometrists
879400	Northwest Orthopedic Spec	Spokane	62111107	Offices of Physicians, Except Mental Health
880400	Optic One Eye Care Ctr	Spokane	62132003	Offices of Optometrists
883400	Pregnancy Help	Spokane	62149301	Freestanding Emergency Medical Centers
884400	Alan D Purdy MD	Spokane	62111107	Offices of Physicians, Except Mental Health
885400	Riverview Lutheran Care Ctr	Spokane	62311016	Nursing Care Facilities
887400	Roth Medical Clinic	Spokane	62111107	Offices of Physicians, Except Mental Health
889400	Roger E Rudd DDS	Spokane	62121003	Offices of Dentists
892400	Spokane Cardiology	Spokane	62111107	Offices of Physicians, Except Mental Health
893400	Spokane Digestive Disease Ctr	Spokane	62111107	Offices of Physicians, Except Mental Health
894400	Spokane Imaging LLC	Spokane	62111107	Offices of Physicians, Except Mental Health
896400	Spokane Orthopaedic & Fracture	Spokane	62111107	Offices of Physicians, Except Mental Health
897400	Spokane Plastic Surgeons	Spokane	62111107	Offices of Physicians, Except Mental Health
898400	Spokane Spinal Ctr	Spokane	62131002	Offices of Chiropractors
899400	Spokane Sports & Physical	Spokane	62134007	Offices of Specialty Therapists
900400	Spokane Urology	Spokane	62111107	Offices of Physicians, Except Mental Health
901400	Surgical Specialists	Spokane	62111107	Offices of Physicians, Except Mental Health
902400	Teen-Aid Inc.	Spokane	62141005	Family Planning Centers
903400	Thera Sport Northwest	Spokane	62134007	Offices of Specialty Therapists
905400	Vanishing Veins	Spokane	62111107	Offices of Physicians, Except Mental Health
907400	Accellaration Physical Thrpy	Spokane	62134007	Offices of Specialty Therapists
908400	Access Endodonic Specialist	Spokane	62149301	Freestanding Emergency Medical Centers

Appendix D cont'd

DesID	NAME	City	NAICS_CODE	NAICS_DESC
909400	Acorn Foot Clinic	Spokane	62139103	Offices of Podiatrists
910400	Steven D Aeschliman DDS	Spokane	62121003	Offices of Dentists
912400	Alterra Clare Bridge	Spokane	62311016	Nursing Care Facilities
916400	Apria Healthcare	Spokane	62199921	Misc Ambulatory Health Care Svcs
917400	Jack Ashlock DDS	Spokane	62121003	Offices of Dentists
918400	Aspen Dental	Spokane	62121003	Offices of Dentists
919400	Associates for Women's Health	Spokane	62111107	Offices of Physicians, Except Mental Health
920400	B & B Physical Therapy	Spokane	62134007	Offices of Specialty Therapists
921400	Bright Now! Dental Ctr	Spokane	62121003	Offices of Dentists
922400	Stephen I Campbell DDS	Spokane	62121003	Offices of Dentists
924400	Cascade Dental Care	Spokane	62121003	Offices of Dentists
925400	Cascade Oral Surgery	Spokane	62121003	Offices of Dentists
926400	Case Care	Spokane	62139920	Offices of Misc Health Practitioners
927400	Center for Advanced Dental	Spokane	62121003	Offices of Dentists
929400	Condon & Condon	Spokane	62121003	Offices of Dentists
930400	Correll & Correll	Spokane	62121003	Offices of Dentists
931400	Randal Phil Defelice MD	Spokane	62111107	Offices of Physicians, Except Mental Health
932400	Dermatology Associates-Spokane	Spokane	62111107	Offices of Physicians, Except Mental Health
933400	Diakonia Family Health	Spokane	62199921	Misc Ambulatory Health Care Svcs
935400	DSI	Spokane	62149301	Freestanding Emergency Medical Centers
936400	Eldercare Inc	Spokane	62311016	Nursing Care Facilities
938400	Engen Engen & Hahn	Spokane	62121003	Offices of Dentists
939400	Englund Dentistry	Spokane	62121003	Offices of Dentists
940400	Raymond Erickson DDS	Spokane	62121003	Offices of Dentists
942400	First Care	Spokane	62111107	Offices of Physicians, Except Mental Health
943400	Franklin Hills Health & Rehab	Spokane	62134007	Offices of Specialty Therapists
944400	Franklin Park Urgent Care Ctr	Spokane	62111107	Offices of Physicians, Except Mental Health
946400	Geriatric Family Care	Spokane	62111107	Offices of Physicians, Except Mental Health

APPENDIX

Appendix D cont'd

DesID	NAME	City	NAICS_CODE	NAICS_DESC
947400	Katherine M Hakes DDS	Spokane	62121003	Offices of Dentists
948400	Foster V Hall DDS	Spokane	62121003	Offices of Dentists
952400	Holy Family Emergency Ctr	Spokane	62111107	Offices of Physicians, Except Mental Health
953400	Holy Family Hosp Cancer Ctr	Spokane	62111107	Offices of Physicians, Except Mental Health
954400	Holy Family Hospital	Spokane	62139925	Offices of Misc Health Practitioners
955400	Holy Family Imaging Ctr	Spokane	62111107	Offices of Physicians, Except Mental Health
956400	Holy Family Multiple Sclr Ctr	Spokane	62111107	Offices of Physicians, Except Mental Health
957400	Holy Family Sleep Disorders	Spokane	62111107	Offices of Physicians, Except Mental Health
958400	Homewatch Care Givers	Spokane	62161001	Home Health Care Svcs
959400	Horizon Hospice	Spokane	62139920	Offices of Misc Health Practitioners
960400	In Motion Physical Therapy	Spokane	62134007	Offices of Specialty Therapists
961400	Indian Trail Dental Care	Spokane	62121003	Offices of Dentists
962400	Indian Trail Family Medicine	Spokane	62111107	Offices of Physicians, Except Mental Health
963400	Inland Eye Ctr	Spokane	62111107	Offices of Physicians, Except Mental Health
964400	Inland Imaging Holy Family	Spokane	62111107	Offices of Physicians, Except Mental Health
965400	Inland Imaging North	Spokane	62111107	Offices of Physicians, Except Mental Health
969400	Kevin King DDS	Spokane	62121003	Offices of Dentists
970400	Dale Paul Lewis DO	Spokane	62111107	Offices of Physicians, Except Mental Health
971400	Lidgerwood Health Care Ctr	Spokane	62111107	Offices of Physicians, Except Mental Health
972400	Lighthouse Counseling	Spokane	62139936	Offices of Misc Health Practitioners
974400	William R Loomis DO	Spokane	62111107	Offices of Physicians, Except Mental Health
975400	Douglas W Lyman OD	Spokane	62132003	Offices of Optometrists
978400	Maternity Support Svc	Spokane	62211002	General Medical & Surgical Hospitals
979400	Microneurosurgery & Spine	Spokane	62111107	Offices of Physicians, Except Mental Health
980400	Miller Chiropractic	Spokane	62131002	Offices of Chiropractors
981400	Ronald R Miller DDS	Spokane	62121003	Offices of Dentists
983400	George H Nauert DDS	Spokane	62121003	Offices of Dentists
984400	North Cedar Dental	Spokane	62121003	Offices of Dentists

Appendix D cont'd

DesID	NAME	City	NAICS_CODE	NAICS_DESC
985400	North Market Chiropractic	Spokane	62131002	Offices of Chiropractors
986400	North Spokane Physical Therapy	Spokane	62134007	Offices of Specialty Therapists
989400	Northside Internal Medicine	Spokane	62111107	Offices of Physicians, Except Mental Health
991400	Northside Physical Therapy	Spokane	62134007	Offices of Specialty Therapists
993400	Northwest Spokane Pediatrics	Spokane	62111107	Offices of Physicians, Except Mental Health
994400	Objective Medical Assessments	Spokane	62199928	Misc Ambulatory Health Care Svcs
995400	Olson Pediatrics	Spokane	62111107	Offices of Physicians, Except Mental Health
996400	One-Plus	Spokane	62199928	Misc Ambulatory Health Care Svcs
997400	Randy E Otterholt DDS	Spokane	62121003	Offices of Dentists
999400	Joseph J Pawlusiak DDS	Spokane	62121003	Offices of Dentists
1000400	Pediatric Associates-Spokane	Spokane	62111107	Offices of Physicians, Except Mental Health
1001400	Pediatric Therapy Svc-Spokane	Spokane	62134007	Offices of Specialty Therapists
1002400	Physical Therapy North	Spokane	62134007	Offices of Specialty Therapists
1003400	Pineview Dental	Spokane	62121003	Offices of Dentists
1007400	Providence Physicians Svc	Spokane	62111107	Offices of Physicians, Except Mental Health
1008400	Provident Services Eastern WA	Spokane	62211002	General Medical & Surgical Hospitals
1009400	Psoriasis-Uva Treatment Ctr	Spokane	62111107	Offices of Physicians, Except Mental Health
1010400	QTC Medical Inc	Spokane	62149301	Freestanding Emergency Medical Centers
1011400	Royal Park Care Ctr	Spokane	62311016	Nursing Care Facilities
1013400	Arthur Rudd DDS	Spokane	62121003	Offices of Dentists
1015400	Michael P Sicilia MD	Spokane	62111107	Offices of Physicians, Except Mental Health
1016400	Spokane Dental Emergency Ctr	Spokane	62121003	Offices of Dentists
1017400	Spokane Occupational Therapy	Spokane	62134006	Offices of Specialty Therapists
1018400	Spokane Urban Indian Health	Spokane	62199921	Misc Ambulatory Health Care Svcs
1019400	Sports Therapy & Rehab	Spokane	62134007	Offices of Specialty Therapists
1020400	St Luke's Rehab at Indian Trl	Spokane	62134007	Offices of Specialty Therapists
1021400	Randall K Stephens DDS	Spokane	62121003	Offices of Dentists
1023400	Elizabeth A Tomeo MD	Spokane	62111107	Offices of Physicians, Except Mental Health

APPENDIX

Appendix D cont'd

DesID	NAME	City	NAICS_CODE	NAICS_DESC
1024400	Christopher Tullis MD	Spokane	62111107	Offices of Physicians, Except Mental Health
1025400	Valente Chiropractic	Spokane	62131002	Offices of Chiropractors
1030400	Affordable Chiropractic	Spokane Valley	62131002	Offices of Chiropractors
1031400	Affordable Dentures	Spokane Valley	62121003	Offices of Dentists
1035400	Cancer Care Northwest Ps	Spokane Valley	62111107	Offices of Physicians, Except Mental Health
1038400	Chiropractic Plus	Spokane Valley	62131002	Offices of Chiropractors
1039400	Chiropractic Works	Spokane Valley	62131002	Offices of Chiropractors
1040400	Civil Air Patrol	Spokane	62191004	Ambulance Svcs
1041400	Gilbert R Escandon MD	Spokane	62111107	Offices of Physicians, Except Mental Health
1042400	Gentiva Health Svc	Spokane Valley	62134007	Offices of Specialty Therapists
1043400	Hillary's Health	Spokane Valley	62111107	Offices of Physicians, Except Mental Health
1044400	Home Care of Washington Inc	Spokane Valley	62161001	Home Health Care Svcs
1050400	Metro Aviation Inc.	Spokane	62149301	Freestanding Emergency Medical Centers
1051400	Millwood Family Dental	Spokane	62121003	Offices of Dentists
1052400	Momentum Physical Therapy	Spokane	62134007	Offices of Specialty Therapists
1053400	Northwest Counseling Group	Spokane Valley	62221001	Psychiatric & Substance Abuse Hospitals
1054400	Northwest Med Star	Spokane	62191002	Ambulance Svcs
1055400	Northwest Therapy Resources	Spokane Valley	62134006	Offices of Specialty Therapists
1058400	Peerless Dental Clinic & Lab	Spokane Valley	62121003	Offices of Dentists
1059400	Progressive Nutrition	Spokane Valley	62139925	Offices of Misc Health Practitioners
1060400	S & S Health Care	Spokane Valley	62139920	Offices of Misc Health Practitioners
1062400	Sir Lawrence Enterprises Inc	Spokane Valley	62121003	Offices of Dentists
1063400	Soul Odyssey	Spokane	62199927	Misc Ambulatory Health Care Svcs
1064400	Spokane Endodontics	Spokane Valley	62121003	Offices of Dentists
1067400	V Care Health Systems Inc	Spokane Valley	62199921	Misc Ambulatory Health Care Svcs
1071400	Zografos Chiropractic Ctr	Spokane Valley	62131002	Offices of Chiropractors
1073400	American Behavioral Health	Spokane Valley	62221001	Psychiatric & Substance Abuse Hospitals
1074400	American Drug Testing	Spokane Valley	62151103	Medical Laboratories

Appendix D cont'd

DesID	NAME	City	NAICS_CODE	NAICS_DESC
1075400	John Arnold PHD	Spokane Valley	62139936	Offices of Misc Health Practitioners
1076400	Aspen Sleep Ctr	Spokane Valley	62139936	Offices of Misc Health Practitioners
1077400	Birthright	Spokane Valley	62139916	Offices of Misc Health Practitioners
1078400	Bourekis & Psomas	Spokane Valley	62121003	Offices of Dentists
1079400	Robert C Brewster MD	Spokane Valley	62111107	Offices of Physicians, Except Mental Health
1081400	Bristles for Kids	Spokane Valley	62121003	Offices of Dentists
1082400	Cancer Care Ctr	Spokane Valley	62111107	Offices of Physicians, Except Mental Health
1084400	Centennial Sports & Physical	Spokane Valley	62134007	Offices of Specialty Therapists
1085400	Children's Dental Village	Spokane Valley	62121003	Offices of Dentists
1087400	Childrens Chiropractic Ctr	Spokane Valley	62131002	Offices of Chiropractors
1091400	Tom O Conlon DDS	Spokane Valley	62121003	Offices of Dentists
1092400	Steven R Daehlin DDS	Spokane Valley	62121003	Offices of Dentists
1093400	Damon & Magnuson	Spokane Valley	62121003	Offices of Dentists
1096400	DSI Spokane Valley Dialysis	Spokane Valley	62149301	Freestanding Emergency Medical Centers
1097400	Ellingsen Endodontics	Spokane Valley	62121003	Offices of Dentists
1098400	Ellingsen Flynn Fmly Dentistry	Spokane Valley	62121003	Offices of Dentists
1099400	Empire Eye Physicians	Spokane Valley	62111107	Offices of Physicians, Except Mental Health
1100400	Empire Eye Surgery Ctr	Spokane Valley	62111107	Offices of Physicians, Except Mental Health
1101400	Evergreen Cosmetic Dentistry	Spokane Valley	62121003	Offices of Dentists
1102400	Evergreen Physical Therapy	Spokane Valley	62134007	Offices of Specialty Therapists
1103400	Gilman & Curalli	Spokane Valley	62139923	Offices of Misc Health Practitioners
1104400	Global Drug Testing Labs Inc	Spokane Valley	62151103	Medical Laboratories
1105400	Robert J Golden MD	Spokane Valley	62111107	Offices of Physicians, Except Mental Health
1106400	Gray Clinical Research	Spokane Valley	62111107	Offices of Physicians, Except Mental Health
1107400	Harken Dental	Spokane Valley	62121003	Offices of Dentists
1108400	Linda S Harrison MD	Spokane Valley	62111107	Offices of Physicians, Except Mental Health
1109400	Healthquest Chiropractic	Spokane Valley	62131002	Offices of Chiropractors
1111400	Higuchi & Skinner	Spokane Valley	62121003	Offices of Dentists

APPENDIX

Appendix D cont'd

DesID	NAME	City	NAICS_CODE	NAICS_DESC
1113400	James A Howard DDS	Spokane Valley	62121003	Offices of Dentists
1114400	Incyte Pathology PS	Spokane Valley	62151106	Medical Laboratories
1115400	Inland Imaging Valley	Spokane Valley	62111107	Offices of Physicians, Except Mental Health
1116400	Inland Northwest Family Foot	Spokane Valley	62111107	Offices of Physicians, Except Mental Health
1117400	Inland Northwest Orthodontic	Spokane Valley	62121003	Offices of Dentists
1118400	Marielle Kwon OD	Spokane Valley	62132003	Offices of Optometrists
1119400	Lake City Physical Therapy	Spokane Valley	62134007	Offices of Specialty Therapists
1120400	Lakeside Physical Therapy	Spokane Valley	62134007	Offices of Specialty Therapists
1121400	Laser Institute at Empire Eye	Spokane Valley	62111107	Offices of Physicians, Except Mental Health
1122400	Tracy Magnuson MD	Spokane Valley	62111107	Offices of Physicians, Except Mental Health
1123400	Mission Avenue Dental Clinic	Spokane Valley	62121003	Offices of Dentists
1125400	Lauralee Nygaard DDS	Spokane Valley	62121003	Offices of Dentists
1127400	Maryam Parviz MD	Spokane Valley	62111107	Offices of Physicians, Except Mental Health
1128400	Pathology Associates	Spokane Valley	62151106	Medical Laboratories
1129400	Pathology Associates Medical	Spokane Valley	62151106	Medical Laboratories
1130400	Pathology Associates Medical	Spokane Valley	62151106	Medical Laboratories
1131400	Performance Physical Therapy	Spokane Valley	62134007	Offices of Specialty Therapists
1132400	Physicians Clinic of Spokane	Spokane Valley	62111107	Offices of Physicians, Except Mental Health
1133400	Pure Health Solutions Inc	Spokane Valley	62199921	Misc Ambulatory Health Care Svcs
1134400	Paul F Reamer DDS	Spokane Valley	62121003	Offices of Dentists
1135400	Rockwood Clinic	Spokane Valley	62149301	Freestanding Emergency Medical Centers
1136400	Rockwood Valley Surgery	Spokane Valley	62111107	Offices of Physicians, Except Mental Health
1137400	Rose Pointe Assisted Living	Spokane Valley	62311016	Nursing Care Facilities
1138400	Ryan R Love Pc	Spokane Valley	62121003	Offices of Dentists
1139400	Seabiotics Distributor	Spokane Valley	62199926	Misc Ambulatory Health Care Svcs
1141400	Senior Helpers	Spokane Valley	62161001	Home Health Care Svcs
1142400	Spokane Cardiology	Spokane Valley	62111107	Offices of Physicians, Except Mental Health
1143400	Spokane Internal Medicine	Spokane Valley	62111107	Offices of Physicians, Except Mental Health

Appendix D cont'd

DesID	NAME	City	NAICS_CODE	NAICS_DESC
1144400	Spokane Obstetrics & Gyn	Spokane Valley	62111107	Offices of Physicians, Except Mental Health
1145400	Spokane Urology	Spokane Valley	62111107	Offices of Physicians, Except Mental Health
1146400	Spokane Valley Family Medicine	Spokane Valley	62111107	Offices of Physicians, Except Mental Health
1147400	St Lukes Rehab in Valley	Spokane Valley	62134006	Offices of Specialty Therapists
1149400	Star Sports Therapy & Rehab	Spokane Valley	62134007	Offices of Specialty Therapists
1150400	Sullivan Park Assisted Living	Spokane Valley	62311016	Nursing Care Facilities
1151400	Sullivan Park Care Ctr	Spokane Valley	62311016	Nursing Care Facilities
1152400	Summit Rehabilitation Assoc	Spokane Valley	62134007	Offices of Specialty Therapists
1153400	Surgical Specialists Spokane	Spokane Valley	62111107	Offices of Physicians, Except Mental Health
1154400	Valley Family Physicians	Spokane Valley	62111107	Offices of Physicians, Except Mental Health
1155400	Valley Family Practice Assoc	Spokane Valley	62111107	Offices of Physicians, Except Mental Health
1156400	Valley Gastroenterology	Spokane Valley	62111107	Offices of Physicians, Except Mental Health
1158400	Valley Hospital & Medical Ctr	Spokane Valley	62111107	Offices of Physicians, Except Mental Health
1160400	Valley Obstetrics & Gyn	Spokane Valley	62111107	Offices of Physicians, Except Mental Health
1161400	Valley Orthopedic Clinic	Spokane Valley	62111107	Offices of Physicians, Except Mental Health
1162400	Valley Outpatient Surgery Ctr	Spokane Valley	62211002	General Medical & Surgical Hospitals
1163400	Valley Rehab & Emg	Spokane Valley	62111107	Offices of Physicians, Except Mental Health
1164400	Valley Rockwood Clinic	Spokane Valley	62111107	Offices of Physicians, Except Mental Health
1166400	Valley Young Peoples Clinic	Spokane Valley	62111107	Offices of Physicians, Except Mental Health
1167400	Veradale Health Care Ctr	Spokane Valley	62111107	Offices of Physicians, Except Mental Health
1168400	Vercler Medical Ctr	Spokane Valley	62111107	Offices of Physicians, Except Mental Health
1170400	Weiand & Weiand	Spokane Valley	62121003	Offices of Dentists
1171400	Sue Weishaar DDS	Spokane Valley	62121003	Offices of Dentists
1173400	Whitworth Physical Therapy	Spokane Valley	62134007	Offices of Specialty Therapists
1174400	Brent M Williams DDS	Spokane Valley	62121003	Offices of Dentists
1175400	Don Williams DO	Spokane Valley	62111107	Offices of Physicians, Except Mental Health
1176400	Youthful Horizons	Spokane Valley	62134006	Offices of Specialty Therapists
1178400	Country Manor	Spokane	62311016	Nursing Care Facilities

APPENDIX

Appendix D cont'd

DesID	NAME	City	NAICS_CODE	NAICS_DESC
1179400	Cudmore Adult Family Home	Spokane	62311016	Nursing Care Facilities
1182400	Elder Services	Spokane	62221001	Psychiatric & Substance Abuse Hospitals
1186400	Spokane Medical Personnel	Spokane	62139920	Offices of Misc Health Practitioners
1187400	A Brief Counseling Ctr	Spokane	62139936	Offices of Misc Health Practitioners
1188400	Advanced Health Chiropractic	Spokane	62131002	Offices of Chiropractors
1189400	Alpine Family Chiropractic Ctr	Spokane	62149301	Freestanding Emergency Medical Centers
1190400	American Mobile Drug Testing	Spokane	62151103	Medical Laboratories
1191400	Apex Physical Therapy	Spokane	62134007	Offices of Specialty Therapists
1193400	Avalon Care Ctr	Spokane	62311016	Nursing Care Facilities
1194400	Beverly Health Ctr	Spokane	62199921	Misc Ambulatory Health Care Svcs
1195400	Laurie S Blevins MD	Spokane	62111107	Offices of Physicians, Except Mental Health
1196400	Braun Dental Care	Spokane	62121003	Offices of Dentists
1197400	A Douglas Brossoit DDS	Spokane	62121003	Offices of Dentists
1198400	Cancer Care Northwest	Spokane	62111107	Offices of Physicians, Except Mental Health
1199400	Chaffin Dental Care	Spokane	62121003	Offices of Dentists
1200400	Children's Choice	Spokane	62121003	Offices of Dentists
1201400	Donald F Condon MD	Spokane	62111107	Offices of Physicians, Except Mental Health
1202400	Country Homes Medical Ctr	Spokane	62111107	Offices of Physicians, Except Mental Health
1203400	Dorosh Dental	Spokane	62121003	Offices of Dentists
1205400	Ronald H Ellingsen DDS	Spokane	62121003	Offices of Dentists
1208400	Laura Fischer OD	Spokane	62132003	Offices of Optometrists
1209400	Foundations of Health	Spokane	62199921	Misc Ambulatory Health Care Svcs
1211400	Donald K Grim DPM	Spokane	62139103	Offices of Podiatrists
1212400	Grummons Orthodontics	Spokane	62121003	Offices of Dentists
1213400	Harmony Family Dental	Spokane	62121003	Offices of Dentists
1215400	Healthy Focus Family Medicine	Spokane	62111107	Offices of Physicians, Except Mental Health
1216400	Robb Heinrich DDS	Spokane	62121003	Offices of Dentists
1217400	Holy Family Physical Therapy	Spokane	62134007	Offices of Specialty Therapists

Appendix D cont'd

DesID	NAME	City	NAICS_CODE	NAICS_DESC
1218400	Holy Family Rhb at Northpoint	Spokane	62111107	Offices of Physicians, Except Mental Health
1219400	Robert M Hughes DDS	Spokane	62121003	Offices of Dentists
1221400	Inland Cardiology Assoc	Spokane	62111107	Offices of Physicians, Except Mental Health
1222400	Inland Northwest Orthodontics	Spokane	62121003	Offices of Dentists
1223400	Institute for Physical & Sprts	Spokane	62134007	Offices of Specialty Therapists
1224400	Jamison Madsen Family Med	Spokane	62111107	Offices of Physicians, Except Mental Health
1225400	Johnson Orthodontics	Spokane	62121003	Offices of Dentists
1226400	Kidds Place Dentistry-Children	Spokane	62121003	Offices of Dentists
1228400	Kinetic Chiropractic	Spokane	62131002	Offices of Chiropractors
1232400	My Dentist	Spokane	62121003	Offices of Dentists
1234400	North Point Dental Arts	Spokane	62121003	Offices of Dentists
1235400	North Point Family Dentistry	Spokane	62121003	Offices of Dentists
1236400	North Spokane Natural Med Ct	Spokane	62131002	Offices of Chiropractors
1237400	Northpointe Dialysis Unit	Spokane	62149202	Kidney Dialysis Centers
1238400	Northside Dentistry	Spokane	62121003	Offices of Dentists
1239400	Northwest Dermatology	Spokane	62111107	Offices of Physicians, Except Mental Health
1240400	Northwest Physical Therapy	Spokane	62134007	Offices of Specialty Therapists
1241400	NPD Svc	Spokane	62121003	Offices of Dentists
1242400	Jeffrey R O'Connor MD	Spokane	62111107	Offices of Physicians, Except Mental Health
1243400	Orthopaedic Specialty Clinic	Spokane	62111107	Offices of Physicians, Except Mental Health
1244400	David Pasino	Spokane	62134007	Offices of Specialty Therapists
1245400	Pathology Associates Medical	Spokane	62151106	Medical Laboratories
1246400	Regency at Northpointe	Spokane	62134007	Offices of Specialty Therapists
1247400	Rockwood at Hawthorne Cmmnty	Spokane	62311016	Nursing Care Facilities
1248400	Rockwood Clinic North	Spokane	62111107	Offices of Physicians, Except Mental Health
1249400	Paul N Ross DDS	Spokane	62121003	Offices of Dentists
1250400	Daniel R Roth DDS	Spokane	62121003	Offices of Dentists
1251400	Douglas G Rundle DDS	Spokane	62121003	Offices of Dentists

APPENDIX

Appendix D cont'd

DesID	NAME	City	NAICS_CODE	NAICS_DESC
1253400	Slack & Combs Orthodontics	Spokane	62121003	Offices of Dentists
1254400	Spokane Back & Neck Clinic	Spokane	62131002	Offices of Chiropractors
1255400	Spokane Chiropractic & Sports	Spokane	62131002	Offices of Chiropractors
1256400	Spokane Dental Arts	Spokane	62121003	Offices of Dentists
1257400	Spokane Ears Nose Throat	Spokane	62111107	Offices of Physicians, Except Mental Health
1259400	Spokane Sports & Spine	Spokane	62111107	Offices of Physicians, Except Mental Health
1260400	Suncrest Wellness Ctr	Spokane	62139923	Offices of Misc Health Practitioners
1261400	Timothy Sweatman DDS	Spokane	62121003	Offices of Dentists
1263400	Tender Touch	Spokane	62199921	Misc Ambulatory Health Care Svcs
1264400	J Boyd Vereen DO	Spokane	62111107	Offices of Physicians, Except Mental Health
1265400	Wandermere Family Dentistry	Spokane	62121003	Offices of Dentists
1268400	Whitworth Physical Therapy	Spokane	62134007	Offices of Specialty Therapists
1270400	Women's Health Connection	Spokane	62111107	Offices of Physicians, Except Mental Health
1271400	Richard M Yarbro DDS	Spokane	62121003	Offices of Dentists
1273400	Advantage Physical Therapy	Spokane	62134007	Offices of Specialty Therapists
1275400	Bryan D Anderson DDS	Spokane	62121003	Offices of Dentists
1276400	Applause Hand Therapy	Spokane	62134006	Offices of Specialty Therapists
1277400	Back & Neck Pain Specialist	Spokane	62131002	Offices of Chiropractors
1278400	Hope M Busto-Keyes	Spokane	62139923	Offices of Misc Health Practitioners
1280400	Cascade Dental Care	Spokane	62121003	Offices of Dentists
1281400	Center for Physical & Sports	Spokane	62134007	Offices of Specialty Therapists
1284400	Collins Family Dentistry	Spokane	62121003	Offices of Dentists
1285400	Steven G Crump DDS	Spokane	62121003	Offices of Dentists
1286400	Dental Care of Spokane	Spokane	62121003	Offices of Dentists
1287400	Dental Clinique	Spokane	62121003	Offices of Dentists
1289400	Bryan K Finn DDS	Spokane	62121003	Offices of Dentists
1290400	First Care	Spokane	62149302	Freestanding Emergency Medical Centers
1294400	Michael P Harwood DDS	Spokane	62121003	Offices of Dentists

Appendix D cont'd

DesID	NAME	City	NAICS_CODE	NAICS_DESC
1295400	James C Hoppe DDS	Spokane	62121003	Offices of Dentists
1296400	Inland Endodontics	Spokane	62121003	Offices of Dentists
1297400	Inland Northwest Genetics Clnc	Spokane	62111107	Offices of Physicians, Except Mental Health
1300400	Larry Ham Physical Therapy Ps	Spokane	62134007	Offices of Specialty Therapists
1301400	Lincoln Heights Dental Ctr	Spokane	62121003	Offices of Dentists
1302400	Susan Mahan-Kohls DDS	Spokane	62121003	Offices of Dentists
1303400	Marineau Healthcare Conslnt	Spokane	62199921	Misc Ambulatory Health Care Svcs
1304400	Ann-Marie Monson DDS	Spokane	62121003	Offices of Dentists
1307400	Mountain View Dental Care	Spokane	62121003	Offices of Dentists
1308400	Kimberly Murdoch DDS	Spokane	62121003	Offices of Dentists
1309400	Nancy Rider MSNARNPCS	Spokane	62199921	Misc Ambulatory Health Care Svcs
1310400	Northwest Physical Therapy Ctr	Spokane	62134007	Offices of Specialty Therapists
1311400	Northwest Pulmonary	Spokane	62111107	Offices of Physicians, Except Mental Health
1312400	Office Anesthesia & Dental	Spokane	62121003	Offices of Dentists
1314400	Postlethwaite Chiropractic	Spokane	62131002	Offices of Chiropractors
1316400	Rockwood Clinic	Spokane	62111107	Offices of Physicians, Except Mental Health
1317400	Rockwood Courtyard	Spokane	62311016	Nursing Care Facilities
1318400	Todd M Rogers DDS	Spokane	62121003	Offices of Dentists
1320400	Serenity Rose	Spokane	62161001	Home Health Care Svcs
1321400	Robert R Shaw DDS	Spokane	62121003	Offices of Dentists
1322400	South Hill Gentle Dentistry	Spokane	62121003	Offices of Dentists
1323400	South Hills Family Medicine	Spokane	62111107	Offices of Physicians, Except Mental Health
1324400	South Regal Health Care Ctr	Spokane	62111107	Offices of Physicians, Except Mental Health
1325400	Southill Periodontics	Spokane	62121003	Offices of Dentists
1327400	Spokane Foot Clinic	Spokane	62139103	Offices of Podiatrists
1328400	Tataryn Endodontics	Spokane	62121003	Offices of Dentists
1330400	Richard Weigand DDS	Spokane	62121003	Offices of Dentists
1331400	Witter & Witter	Spokane	62121003	Offices of Dentists

APPENDIX

Appendix D cont'd

DesID	NAME	City	NAICS_CODE	NAICS_DESC
1332400	Ronald Yep MD	Spokane	62111107	Offices of Physicians, Except Mental Health
1334400	Alliance for Affordable Svc	Spokane	62199921	Misc Ambulatory Health Care Svcs
1339400	Meadowbrook Educational Svc	Spokane	62199956	Misc Ambulatory Health Care Svcs
1342400	Vista Consulting Svc	Spokane	62134005	Offices of Specialty Therapists
1343400	West Plains Chiropractic	Airway Heights	62131002	Offices of Chiropractors
1344400	Lynne H Williams MD	Colbert	62111107	Offices of Physicians, Except Mental Health
1345400	Deer Park Hospital	Deer Park	62134007	Offices of Specialty Therapists
1347400	General Surgey Clinic	Fairchild Afb	62111107	Offices of Physicians, Except Mental Health
1350400	Macdonald Chiropractic	Mead	62131002	Offices of Chiropractors
1351400	Maria Lydia Montenegro MD	Medical Lake	62111107	Offices of Physicians, Except Mental Health
1352400	Kathryn Moore	Medical Lake	62139932	Offices of Misc Health Practitioners
1353400	Bonhams Adult Family Home	Nine Mile Falls	62311006	Nursing Care Facilities
1355400	Phase 1 Physical Therapy	Nine Mile Falls	62134007	Offices of Specialty Therapists
1356400	Holistic Physical Therapy	Spokane Valley	62134007	Offices of Specialty Therapists
1357400	Evergreen Ultrasound	Spokane	62111107	Offices of Physicians, Except Mental Health
1359400	Outreach Therapy Conslnts Inc	Spokane	62134006	Offices of Specialty Therapists
1361400	Planned Parenthood	Spokane	62141001	Family Planning Centers
1365400	Nick Wissink OD	Spokane	62132003	Offices of Optometrists
1367400	Deaconess Medical Ctr	Spokane	62111107	Offices of Physicians, Except Mental Health
1368400	Deaconess Women's Clinic	Spokane	62111107	Offices of Physicians, Except Mental Health
1369400	Hospice of Spokane	Spokane	62139920	Offices of Misc Health Practitioners
1370400	C Harold Mielke Jr MD	Spokane	62111107	Offices of Physicians, Except Mental Health
1371400	Northwest Telehealth	Spokane	62199921	Misc Ambulatory Health Care Svcs
1373400	Shriners Hospital	Spokane	62211002	General Medical & Surgical Hospitals
1376400	John C Wantulok DDS	Spokane Valley	62121003	Offices of Dentists
1377400	First Choice Health Care Svc	Spokane Valley	62161001	Home Health Care Svcs
1379400	Spencer A Saito DDS	Spokane Valley	62121003	Offices of Dentists
1380400	Spokane Emergency Physicians	Spokane Valley	62149301	Freestanding Emergency Medical Centers

Appendix D cont'd

DesID	NAME	City	NAICS_CODE	NAICS_DESC
1384400	Lilac City Physical Therapy	Spokane	62134007	Offices of Specialty Therapists
1388400	Pathology Associates Medical	Spokane	62151106	Medical Laboratories
1389400	Sacred Heart Med Laboratory	Spokane	62111107	Offices of Physicians, Except Mental Health
1392400	Cornelis B Bakker MD	Spokane	62111107	Offices of Physicians, Except Mental Health
1393400	First Care	Spokane	62149302	Freestanding Emergency Medical Centers
1395400	Foot Clinic	Spokane	62111107	Offices of Physicians, Except Mental Health
1396400	Russell Kelley	Spokane	62139923	Offices of Misc Health Practitioners
1397400	Lifecycle Personal Care Agency	Spokane	62161001	Home Health Care Svcs
1398400	Osdat Llc	Spokane	62151103	Medical Laboratories
1399400	Robert J Rose MD	Spokane	62111107	Offices of Physicians, Except Mental Health
1401400	Riverside Dental Clinic	Chattaroy	62121003	Offices of Dentists
1402400	Smith Orthodontics	Chattaroy	62121003	Offices of Dentists
1404400	Mt Spokane Physical Therapy	Mead	62134007	Offices of Specialty Therapists
1407400	Center for Physical & Sports	Spokane	62134007	Offices of Specialty Therapists
1408400	Thomas J Hansen MD	Spokane	62111107	Offices of Physicians, Except Mental Health
1409400	North Spokane Chiropractic	Spokane	62131002	Offices of Chiropractors
1410400	U S Healthworks Medical Clinic	Spokane	62149301	Freestanding Emergency Medical Centers
1411400	Robert Ulland OD	Spokane	62132003	Offices of Optometrists
1413400	Airway Heights Physical Thrpy	Spokane	62134007	Offices of Specialty Therapists
1414400	Dr David L Turner Dmd	Spokane	62121003	Offices of Dentists

Appendix E: Demographic Data of Focus Census Blocks

GEOID10	NAME10	POP10	GEOID10	NAME10	POP10
530630111022009	Block 2009	153	530630009001009	Block 1009	63
530630032002008	Block 2008	0	530630024001122	Block 1122	0
530630031001001	Block 1001	368	530630031001019	Block 1019	74
530630002003001	Block 3001	28	530630016002003	Block 2003	24
530630002003022	Block 3022	0	530630024001002	Block 1002	3
530630002003024	Block 3024	26	530630024001006	Block 1006	6
530630002003027	Block 3027	6	530630035002078	Block 2078	6
530630140013002	Block 3002	109	530630035002002	Block 2002	0
530630020004005	Block 4005	17	530630035002034	Block 2034	0
530630016002008	Block 2008	30	530630035002027	Block 2027	6
530630104011014	Block 1014	658	530630035002039	Block 2039	0
530630024001055	Block 1055	0	530630032002007	Block 2007	147
530630024001038	Block 1038	0	530630023001033	Block 1033	23
530630024002040	Block 2040	9	530630024001086	Block 1086	4
530630104011021	Block 1021	80	530630020004054	Block 4054	13
530630023001007	Block 1007	21	530630024001066	Block 1066	0
530630023001028	Block 1028	0	530630020004008	Block 4008	12
530630023001036	Block 1036	58	530630016001040	Block 1040	29
530630104011027	Block 1027	41	530630016002029	Block 2029	71
530630024002007	Block 2007	0	530630016001042	Block 1042	73
530630020004016	Block 4016	16	530630016001035	Block 1035	15
530630024001009	Block 1009	2	530630016001026	Block 1026	0
530630020004041	Block 4041	17	530630016001048	Block 1048	7
530630020004040	Block 4040	18	530630023001016	Block 1016	63
530630026003026	Block 3026	15	530630016001001	Block 1001	0
530630024001029	Block 1029	13	530630104011012	Block 1012	0

APPENDIX

Appendix E cont'd

GEOID10	NAME10	POP10	GEOID10	NAME10	POP10
530630026003035	Block 3035	12	530630140013008	Block 3008	0
530630104011041	Block 1041	1	530630032002016	Block 2016	95
530630026003037	Block 3037	20	530630111022006	Block 2006	69
530630020004058	Block 4058	9	530630111022007	Block 2007	263
530630023001014	Block 1014	29	530630111022012	Block 2012	12
530630026003050	Block 3050	18	530630024001041	Block 1041	0
530630024001069	Block 1069	15	530630024001047	Block 1047	0
530630023001026	Block 1026	16	530630024001010	Block 1010	0
530630023001025	Block 1025	31	530630020004011	Block 4011	0
530630024001094	Block 1094	0	530630016002023	Block 2023	6
530630009001006	Block 1006	57	530630140013007	Block 3007	0
530630024002000	Block 2000	3	530630140013000	Block 3000	18
530630032002015	Block 2015	0	530630016002015	Block 2015	12
530630104011015	Block 1015	43	530630032002012	Block 2012	17
530630035002076	Block 2076	49	530630140013004	Block 3004	24
530630035002084	Block 2084	0	530630035002021	Block 2021	0
530630104011001	Block 1001	0	530630035002056	Block 2056	71
530630104011025	Block 1025	70	530630035002028	Block 2028	0
530630140011005	Block 1005	0	530630035002024	Block 2024	37
530630016001050	Block 1050	0	530630035002082	Block 2082	0
530630026003005	Block 3005	0	530630024002013	Block 2013	21
530630016002027	Block 2027	24	530630024002011	Block 2011	15
530630026003007	Block 3007	18	530630024002023	Block 2023	20
530630026003028	Block 3028	20	530630024002022	Block 2022	2
530630026003040	Block 3040	28	530630023001039	Block 1039	21
530630023001037	Block 1037	25	530630020004030	Block 4030	21
530630024001088	Block 1088	19	530630024001106	Block 1106	0
530630140013005	Block 3005	111	530630024001093	Block 1093	43

APPENDIX

Appendix E cont'd

GEOID10	NAME10	POP10	GEOID10	NAME10	POP10
530630035002040	Block 2040	178	530630032002013	Block 2013	0
530630111022011	Block 2011	0	530630032002010	Block 2010	4
530630016001027	Block 1027	0	530630016002010	Block 2010	20
530630002003029	Block 3029	1	530630020004001	Block 4001	15
530630002003038	Block 3038	4	530630020004023	Block 4023	15
530630016001020	Block 1020	417	530630020004033	Block 4033	28
530630023001029	Block 1029	27	530630026003048	Block 3048	16
530630020004056	Block 4056	0	530630024001078	Block 1078	20
530630020004015	Block 4015	15	530630023001035	Block 1035	25
530630024002012	Block 2012	24	530630024001105	Block 1105	36
530630020004019	Block 4019	17	530630035002026	Block 2026	0
530630026003000	Block 3000	7	530630016002007	Block 2007	32
530630026003013	Block 3013	16	530630016002026	Block 2026	43
530630026003021	Block 3021	22	530630016002014	Block 2014	15
530630024002036	Block 2036	26	530630026003004	Block 3004	1
530630026003036	Block 3036	42	530630026003002	Block 3002	122
530630104011000	Block 1000	0	530630026003009	Block 3009	1
530630104011029	Block 1029	0	530630026003031	Block 3031	15
530630104011036	Block 1036	41	530630026003053	Block 3053	22
530630024001098	Block 1098	0	530630002003033	Block 3033	7
530630023001004	Block 1004	19	530630002003041	Block 3041	22
530630026003039	Block 3039	17	530630016001004	Block 1004	49
530630026003046	Block 3046	19	530630016001029	Block 1029	0
530630023001020	Block 1020	25	530630016001051	Block 1051	0
530630024001068	Block 1068	12	530630031001009	Block 1009	85
530630023001042	Block 1042	26	530630002003009	Block 3009	0
530630023001041	Block 1041	35	530630002003012	Block 3012	30
530630035002029	Block 2029	0	530630002003018	Block 3018	0

Appendix E cont'd

GEOID10	NAME10	POP10	GEOID10	NAME10	POP10
530630104011017	Block 1017	52	530630002003023	Block 3023	0
530630035002001	Block 2001	0	530630024002006	Block 2006	84
530630035002036	Block 2036	45	530630020004006	Block 4006	23
530630035002037	Block 2037	67	530630020004020	Block 4020	16
530630035002020	Block 2020	1	530630024001051	Block 1051	0
530630035002038	Block 2038	0	530630020004057	Block 4057	54
530630026003001	Block 3001	0	530630026003051	Block 3051	8
530630026003018	Block 3018	9	530630023001019	Block 1019	30
530630026003055	Block 3055	14	530630024001056	Block 1056	502
530630026003016	Block 3016	28	530630032002001	Block 2001	0
530630024001101	Block 1101	0	530630002003008	Block 3008	5
530630035002004	Block 2004	0	530630016001009	Block 1009	0
530630140013010	Block 3010	67	530630016001008	Block 1008	0
530630140013009	Block 3009	130	530630016001022	Block 1022	19
530630140013019	Block 3019	48	530630024002015	Block 2015	22
530630140011001	Block 1001	33	530630024002033	Block 2033	20
530630140011009	Block 1009	29	530630024001036	Block 1036	0
530630140011008	Block 1008	2	530630024002039	Block 2039	65
530630002003007	Block 3007	25	530630023001027	Block 1027	42
530630002003015	Block 3015	32	530630002003013	Block 3013	32
530630002003016	Block 3016	8	530630035002033	Block 2033	0
530630016001019	Block 1019	0	530630035002048	Block 2048	0
530630002003030	Block 3030	47	530630035002044	Block 2044	0
530630016001006	Block 1006	29	530630035002065	Block 2065	0
530630002003021	Block 3021	0	530630035002070	Block 2070	75
530630016001013	Block 1013	28	530630035002069	Block 2069	0
530630016001010	Block 1010	199	530630035002074	Block 2074	0
530630020004043	Block 4043	16	530630035002075	Block 2075	0

APPENDIX

Appendix E cont'd

GEOID10	NAME10	POP10	GEOID10	NAME10	POP10
530630024001067	Block 1067	0	530630035002088	Block 2088	0
530630140013001	Block 3001	48	530630035002083	Block 2083	0
530630016001016	Block 1016	0	530630035002085	Block 2085	0
530630016001028	Block 1028	0	530630104011003	Block 1003	3
530630016001049	Block 1049	2	530630104011018	Block 1018	48
530630016001052	Block 1052	1	530630104011033	Block 1033	34
530630111011005	Block 1005	330	530630104011007	Block 1007	0
530630020004018	Block 4018	13	530630104011038	Block 1038	0
530630002003032	Block 3032	41	530630009001011	Block 1011	51
530630024001016	Block 1016	21	530630009001008	Block 1008	58
530630024001085	Block 1085	1	530630031001014	Block 1014	42
530630140013012	Block 3012	488	530630016001003	Block 1003	8
530630140013017	Block 3017	33	530630026003019	Block 3019	15
530630024001100	Block 1100	0	530630026003024	Block 3024	20
530630035002018	Block 2018	3	530630026003041	Block 3041	18
530630035002061	Block 2061	0	530630024001080	Block 1080	5
530630035002059	Block 2059	0	530630024001123	Block 1123	0
530630035002055	Block 2055	26	530630024001084	Block 1084	33
530630032002003	Block 2003	71	530630035002051	Block 2051	0
530630035002052	Block 2052	0	530630035002064	Block 2064	0
530630024002003	Block 2003	10	530630104011009	Block 1009	0
530630024002018	Block 2018	29	530630104011032	Block 1032	42
530630024001034	Block 1034	0	530630104011028	Block 1028	34
530630024002024	Block 2024	15	530630104011011	Block 1011	2184
530630024001054	Block 1054	0	530630104011010	Block 1010	0
530630024001026	Block 1026	63	530630104011046	Block 1046	12
530630024001039	Block 1039	0	530630016002030	Block 2030	42
530630024001003	Block 1003	3	530630016001038	Block 1038	12

Appendix E cont'd

GEOID10	NAME10	POP10	GEOID10	NAME10	POP10
530630111022004	Block 2004	0	530630140011004	Block 1004	18
530630032002002	Block 2002	1	530630032002004	Block 2004	92
530630140011002	Block 1002	10	530630035002041	Block 2041	0
530630140013014	Block 3014	15	530630032002000	Block 2000	147
530630140013018	Block 3018	45	530630016001002	Block 1002	14
530630111022001	Block 2001	114	530630002003026	Block 3026	1
530630035002062	Block 2062	0	530630016001023	Block 1023	2
530630035002003	Block 2003	0	530630111011001	Block 1001	0
530630035002010	Block 2010	0	530630031001011	Block 1011	61
530630035002030	Block 2030	104	530630016002004	Block 2004	33
530630035002050	Block 2050	0	530630016002005	Block 2005	30
530630035002072	Block 2072	14	530630016001021	Block 1021	2
530630104011023	Block 1023	5	530630111022013	Block 2013	382
530630104011016	Block 1016	42	530630035002079	Block 2079	124
530630020004044	Block 4044	0	530630104011035	Block 1035	29
530630024001083	Block 1083	0	530630016001043	Block 1043	50
530630031001016	Block 1016	44	530630016001045	Block 1045	52
530630031001010	Block 1010	55	530630016001033	Block 1033	14
530630031001017	Block 1017	31	530630016001032	Block 1032	3
530630035002015	Block 2015	0	530630016001053	Block 1053	7
530630024001052	Block 1052	0	530630016001015	Block 1015	0
530630024001015	Block 1015	0	530630016002020	Block 2020	17
530630024001048	Block 1048	0	530630016001044	Block 1044	66
530630140011007	Block 1007	243	530630111011004	Block 1004	51
530630035002086	Block 2086	0	530630024002008	Block 2008	26
530630024001110	Block 1110	0	530630024002026	Block 2026	22
530630016002016	Block 2016	13	530630111022010	Block 2010	0
530630024001091	Block 1091	0	530630024001111	Block 1111	0

APPENDIX

Appendix E cont'd

GEOID10	NAME10	POP10	GEOID10	NAME10	POP10
530630024001092	Block 1092	0	530630031001008	Block 1008	35
530630009001001	Block 1001	107	530630140013011	Block 3011	108
530630009001005	Block 1005	41	530630035002014	Block 2014	0
530630009001000	Block 1000	78	530630016002018	Block 2018	9
530630024001102	Block 1102	0	530630016001000	Block 1000	17
530630020004013	Block 4013	24	530630016001018	Block 1018	1
530630020004045	Block 4045	8	530630016001017	Block 1017	2
530630020004021	Block 4021	19	530630016001014	Block 1014	0
530630020004024	Block 4024	12	530630016001041	Block 1041	18
530630020004027	Block 4027	17	530630024001119	Block 1119	0
530630020004042	Block 4042	0	530630024001059	Block 1059	0
530630026003015	Block 3015	12	530630035002089	Block 2089	0
530630020004050	Block 4050	20	530630024001049	Block 1049	0
530630024002034	Block 2034	25	530630016001030	Block 1030	0
530630026003025	Block 3025	19	530630024001042	Block 1042	0
530630026003045	Block 3045	16	530630023001003	Block 1003	11
530630023001013	Block 1013	15	530630104011008	Block 1008	0
530630035002058	Block 2058	0	530630035002067	Block 2067	0
530630035002080	Block 2080	0	530630111022005	Block 2005	440
530630035002081	Block 2081	0	530630024001025	Block 1025	0
530630035002073	Block 2073	39	530630024001012	Block 1012	0
530630111011000	Block 1000	551	530630024002014	Block 2014	40
530630020004049	Block 4049	0	530630024002029	Block 2029	44
530630024001076	Block 1076	32	530630024002027	Block 2027	37
530630024001043	Block 1043	1	530630104011006	Block 1006	0
530630020004034	Block 4034	10	530630104011034	Block 1034	40
530630020004052	Block 4052	0	530630016001024	Block 1024	11
530630024001107	Block 1107	0	530630016001034	Block 1034	29

Appendix E cont'd

GEOID10	NAME10	POP10	GEOID10	NAME10	POP10
530630002003039	Block 3039	16	530630016001046	Block 1046	24
530630035002035	Block 2035	0	530630016001025	Block 1025	5
530630016002024	Block 2024	11	530630024002002	Block 2002	0
530630016002017	Block 2017	10	530630024002037	Block 2037	44
530630016002028	Block 2028	51	530630024002035	Block 2035	28
530630023001044	Block 1044	9	530630024001028	Block 1028	4
530630023001043	Block 1043	6	530630020004004	Block 4004	0
530630024001104	Block 1104	0	530630024001096	Block 1096	55
530630035002032	Block 2032	44	530630104011024	Block 1024	0
530630035002031	Block 2031	0	530630104011013	Block 1013	0
530630140013003	Block 3003	23	530630002003019	Block 3019	0
530630024002020	Block 2020	12	530630023001002	Block 1002	0
530630024002019	Block 2019	36	530630024001017	Block 1017	0
530630024001037	Block 1037	0	530630026003010	Block 3010	6
530630024002031	Block 2031	7	530630024001031	Block 1031	0
530630024001040	Block 1040	0	530630104011004	Block 1004	0
530630024001115	Block 1115	0	530630016002000	Block 2000	29
530630024001117	Block 1117	0	530630016002013	Block 2013	19
530630035002005	Block 2005	0	530630026003006	Block 3006	0
530630035002007	Block 2007	0	530630020004010	Block 4010	15
530630024001090	Block 1090	0	530630023001030	Block 1030	25
530630031001005	Block 1005	0	530630020004037	Block 4037	24
530630031001004	Block 1004	0	530630020004022	Block 4022	12
530630024001116	Block 1116	0	530630020004014	Block 4014	13
530630032002009	Block 2009	58	530630020004039	Block 4039	18
530630031001012	Block 1012	111	530630026003014	Block 3014	12
530630035002017	Block 2017	70	530630024002038	Block 2038	0
530630035002077	Block 2077	17	530630023001015	Block 1015	17

APPENDIX

Appendix E cont'd

GEOID10	NAME10	POP10	GEOID10	NAME10	POP10
530630035002057	Block 2057	16	530630024001065	Block 1065	21
530630031001003	Block 1003	0	530630024001064	Block 1064	30
530630104011019	Block 1019	40	530630024001063	Block 1063	11
530630104011020	Block 1020	58	530630026003049	Block 3049	32
530630104011026	Block 1026	55	530630023001021	Block 1021	98
530630140011006	Block 1006	353	530630024001070	Block 1070	34
530630002003021	Block 3021	0	530630035002070	Block 2070	75
530630016001013	Block 1013	28	530630035002069	Block 2069	0
530630016001010	Block 1010	199	530630035002074	Block 2074	0
530630020004043	Block 4043	16	530630035002075	Block 2075	0
530630024001067	Block 1067	0	530630035002088	Block 2088	0
530630140013001	Block 3001	48	530630035002083	Block 2083	0
530630016001016	Block 1016	0	530630035002085	Block 2085	0
530630016001028	Block 1028	0	530630104011003	Block 1003	3
530630016001049	Block 1049	2	530630104011018	Block 1018	48
530630016001052	Block 1052	1	530630104011033	Block 1033	34
530630111011005	Block 1005	330	530630104011007	Block 1007	0
530630020004018	Block 4018	13	530630104011038	Block 1038	0
530630002003032	Block 3032	41	530630009001011	Block 1011	51
530630024001016	Block 1016	21	530630009001008	Block 1008	58
530630024001085	Block 1085	1	530630031001014	Block 1014	42
530630140013012	Block 3012	488	530630016001003	Block 1003	8
530630140013017	Block 3017	33	530630026003019	Block 3019	15
530630024001100	Block 1100	0	530630026003024	Block 3024	20
530630035002018	Block 2018	3	530630026003041	Block 3041	18
530630035002061	Block 2061	0	530630024001080	Block 1080	5
530630035002059	Block 2059	0	530630024001123	Block 1123	0
530630035002055	Block 2055	26	530630024001084	Block 1084	33

APPENDIX

Appendix E cont'd

GEOID10	NAME10	POP10	GEOID10	NAME10	POP10
530630032002003	Block 2003	71	530630035002051	Block 2051	0
530630035002052	Block 2052	0	530630035002064	Block 2064	0
530630024002003	Block 2003	10	530630104011009	Block 1009	0
530630024002018	Block 2018	29	530630104011032	Block 1032	42
530630024001034	Block 1034	0	530630104011028	Block 1028	34
530630024002024	Block 2024	15	530630104011011	Block 1011	2184
530630024001054	Block 1054	0	530630104011010	Block 1010	0
530630024001026	Block 1026	63	530630104011046	Block 1046	12
530630024001039	Block 1039	0	530630016002030	Block 2030	42
530630024001003	Block 1003	3	530630016001038	Block 1038	12
530630111022004	Block 2004	0	530630140011004	Block 1004	18
530630032002002	Block 2002	1	530630032002004	Block 2004	92
530630140011002	Block 1002	10	530630035002041	Block 2041	0
530630140013014	Block 3014	15	530630032002000	Block 2000	147
530630140013018	Block 3018	45	530630016001002	Block 1002	14
530630111022001	Block 2001	114	530630002003026	Block 3026	1
530630035002062	Block 2062	0	530630016001023	Block 1023	2
530630035002003	Block 2003	0	530630111011001	Block 1001	0
530630035002010	Block 2010	0	530630031001011	Block 1011	61
530630035002030	Block 2030	104	530630016002004	Block 2004	33
530630035002050	Block 2050	0	530630016002005	Block 2005	30
530630035002072	Block 2072	14	530630016001021	Block 1021	2
530630104011023	Block 1023	5	530630111022013	Block 2013	382
530630104011016	Block 1016	42	530630035002079	Block 2079	124
530630020004044	Block 4044	0	530630104011035	Block 1035	29
530630024001083	Block 1083	0	530630016001043	Block 1043	50
530630031001016	Block 1016	44	530630016001045	Block 1045	52
530630031001010	Block 1010	55	530630016001033	Block 1033	14

205

APPENDIX

Appendix E cont'd

GEOID10	NAME10	POP10	GEOID10	NAME10	POP10
530630031001017	Block 1017	31	530630016001032	Block 1032	3
530630035002015	Block 2015	0	530630016001053	Block 1053	7
530630024001052	Block 1052	0	530630016001015	Block 1015	0
530630024001015	Block 1015	0	530630016002020	Block 2020	17
530630024001048	Block 1048	0	530630016001044	Block 1044	66
530630140011007	Block 1007	243	530630111011004	Block 1004	51
530630035002086	Block 2086	0	530630024002008	Block 2008	26
530630024001110	Block 1110	0	530630024002026	Block 2026	22
530630016002016	Block 2016	13	530630111022010	Block 2010	0
530630024001091	Block 1091	0	530630024001111	Block 1111	0
530630024001092	Block 1092	0	530630031001008	Block 1008	35
530630009001001	Block 1001	107	530630140013011	Block 3011	108
530630009001005	Block 1005	41	530630035002014	Block 2014	0
530630009001000	Block 1000	78	530630016002018	Block 2018	9
530630024001102	Block 1102	0	530630016001000	Block 1000	17
530630020004013	Block 4013	24	530630016001018	Block 1018	1
530630020004045	Block 4045	8	530630016001017	Block 1017	2
530630020004021	Block 4021	19	530630016001014	Block 1014	0
530630020004024	Block 4024	12	530630016001041	Block 1041	18
530630020004027	Block 4027	17	530630024001119	Block 1119	0
530630020004042	Block 4042	0	530630024001059	Block 1059	0
530630026003015	Block 3015	12	530630035002089	Block 2089	0
530630020004050	Block 4050	20	530630024001049	Block 1049	0
530630024002034	Block 2034	25	530630016001030	Block 1030	0
530630026003025	Block 3025	19	530630024001042	Block 1042	0
530630026003045	Block 3045	16	530630023001003	Block 1003	11
530630023001013	Block 1013	15	530630104011008	Block 1008	0
530630035002058	Block 2058	0	530630035002067	Block 2067	0

Appendix E cont'd

GEOID10	NAME10	POP10	GEOID10	NAME10	POP10
530630035002080	Block 2080	0	530630111022005	Block 2005	440
530630035002081	Block 2081	0	530630024001025	Block 1025	0
530630035002073	Block 2073	39	530630024001012	Block 1012	0
530630111011000	Block 1000	551	530630024002014	Block 2014	40
530630020004049	Block 4049	0	530630024002029	Block 2029	44
530630024001076	Block 1076	32	530630024002027	Block 2027	37
530630024001043	Block 1043	1	530630104011006	Block 1006	0
530630020004034	Block 4034	10	530630104011034	Block 1034	40
530630020004052	Block 4052	0	530630016001024	Block 1024	11
530630024001107	Block 1107	0	530630016001034	Block 1034	29
530630002003039	Block 3039	16	530630016001046	Block 1046	24
530630035002035	Block 2035	0	530630016001025	Block 1025	5
530630016002024	Block 2024	11	530630024002002	Block 2002	0
530630016002017	Block 2017	10	530630024002037	Block 2037	44
530630016002028	Block 2028	51	530630024002035	Block 2035	28
530630023001044	Block 1044	9	530630024001028	Block 1028	4
530630023001043	Block 1043	6	530630020004004	Block 4004	0
530630024001104	Block 1104	0	530630024001096	Block 1096	55
530630035002032	Block 2032	44	530630104011024	Block 1024	0
530630035002031	Block 2031	0	530630104011013	Block 1013	0
530630140013003	Block 3003	23	530630002003019	Block 3019	0
530630024002020	Block 2020	12	530630023001002	Block 1002	0
530630024002019	Block 2019	36	530630024001017	Block 1017	0
530630024001037	Block 1037	0	530630026003010	Block 3010	6
530630024002031	Block 2031	7	530630024001031	Block 1031	0
530630024001040	Block 1040	0	530630104011004	Block 1004	0
530630024001115	Block 1115	0	530630016002000	Block 2000	29
530630024001117	Block 1117	0	530630016002013	Block 2013	19

APPENDIX

Appendix E cont'd

GEOID10	NAME10	POP10	GEOID10	NAME10	POP10
530630035002005	Block 2005	0	530630026003006	Block 3006	0
530630035002007	Block 2007	0	530630020004010	Block 4010	15
530630024001090	Block 1090	0	530630023001030	Block 1030	25
530630031001005	Block 1005	0	530630020004037	Block 4037	24
530630031001004	Block 1004	0	530630020004022	Block 4022	12
530630024001116	Block 1116	0	530630020004014	Block 4014	13
530630032002009	Block 2009	58	530630020004039	Block 4039	18
530630031001012	Block 1012	111	530630026003014	Block 3014	12
530630035002017	Block 2017	70	530630024002038	Block 2038	0
530630035002077	Block 2077	17	530630023001015	Block 1015	17
530630035002057	Block 2057	16	530630024001065	Block 1065	21
530630031001003	Block 1003	0	530630024001064	Block 1064	30
530630104011019	Block 1019	40	530630024001063	Block 1063	11
530630104011020	Block 1020	58	530630026003049	Block 3049	32
530630104011026	Block 1026	55	530630023001021	Block 1021	98
530630140011006	Block 1006	353	530630024001070	Block 1070	34
530630002003000	Block 3000	6	530630024001079	Block 1079	17
530630016002011	Block 2011	23	530630035002025	Block 2025	4
530630016001047	Block 1047	26	530630104011047	Block 1047	5
530630016001031	Block 1031	15	530630104011037	Block 1037	0
530630016001055	Block 1055	19	530630104011040	Block 1040	9
530630016001057	Block 1057	27	530630024001018	Block 1018	2
530630111022003	Block 2003	0	530630024001019	Block 1019	0
530630023001024	Block 1024	12	530630023001040	Block 1040	23
530630024001087	Block 1087	0	530630023001038	Block 1038	24
530630002003036	Block 3036	0	530630023001010	Block 1010	22
530630002003003	Block 3003	63	530630023001017	Block 1017	17
530630140013006	Block 3006	18	530630023001009	Block 1009	8

APPENDIX

Appendix E cont'd

GEOID10	NAME10	POP10	GEOID10	NAME10	POP10
530630140013015	Block 3015	193	530630023001031	Block 1031	21
530630035002016	Block 2016	0	530630002003011	Block 3011	44
530630035002013	Block 2013	0	530630016001036	Block 1036	5
530630031001002	Block 1002	176	530630024001008	Block 1008	0
530630024001074	Block 1074	0	530630024001033	Block 1033	10
530630024001014	Block 1014	0	530630024001108	Block 1108	0
530630104011002	Block 1002	0	530630002003020	Block 3020	0
530630002003004	Block 3004	37	530630024001081	Block 1081	0
530630002003034	Block 3034	0	530630024001114	Block 1114	0
530630002003028	Block 3028	2	530630024001082	Block 1082	54
530630035002049	Block 2049	0	530630111011006	Block 1006	0
530630035002054	Block 2054	0	530630026003032	Block 3032	7
530630026003012	Block 3012	26	530630035002087	Block 2087	0
530630026003033	Block 3033	14	530630002003002	Block 3002	33
530630026003008	Block 3008	13	530630111022000	Block 2000	0
530630026003052	Block 3052	12	530630024001057	Block 1057	0
530630026003020	Block 3020	83	530630035002006	Block 2006	0
530630026003029	Block 3029	36	530630031001018	Block 1018	13
530630002003005	Block 3005	36	530630035002068	Block 2068	0
530630104011043	Block 1043	29	530630035002066	Block 2066	0
530630024001053	Block 1053	0	530630024001099	Block 1099	0
530630024001118	Block 1118	0	530630035002023	Block 2023	187
530630035002012	Block 2012	0	530630035002046	Block 2046	36
530630031001020	Block 1020	30	530630035002060	Block 2060	9
530630016001056	Block 1056	4	530630035002045	Block 2045	0
530630035002063	Block 2063	0	530630035002042	Block 2042	50
530630002003017	Block 3017	0	530630009001003	Block 1003	54
530630031001006	Block 1006	55	530630016002021	Block 2021	325

APPENDIX

Appendix E cont'd

GEOID10	NAME10	POP10	GEOID10	NAME10	POP10
530630031001000	Block 1000	7	530630016002025	Block 2025	2
530630016002002	Block 2002	25	530630035002071	Block 2071	25
530630020004003	Block 4003	7	530630032002014	Block 2014	53
530630020004038	Block 4038	40	530630032002011	Block 2011	39
530630024001071	Block 1071	22	530630026003030	Block 3030	23
530630016002006	Block 2006	21	530630024001004	Block 1004	0
530630002003025	Block 3025	1	530630024001013	Block 1013	57
530630016001012	Block 1012	11	530630035002011	Block 2011	0
530630002003031	Block 3031	35	530630035002008	Block 2008	0
530630002003035	Block 3035	0	530630024001020	Block 1020	0
530630016001007	Block 1007	0	530630020004051	Block 4051	5
530630002003010	Block 3010	56	530630020004029	Block 4029	22
530630002003040	Block 3040	15	530630026003056	Block 3056	9
530630032002005	Block 2005	0	530630002003006	Block 3006	34
530630009001002	Block 1002	44	530630020004031	Block 4031	21
530630035002009	Block 2009	0	530630020004047	Block 4047	0
530630035002000	Block 2000	0	530630020004032	Block 4032	18
530630035002019	Block 2019	0	530630024001061	Block 1061	23
530630035002047	Block 2047	21	530630020004053	Block 4053	0
530630035002043	Block 2043	5	530630024001060	Block 1060	25
530630024002005	Block 2005	12	530630020004000	Block 4000	24
530630020004007	Block 4007	8	530630016001005	Block 1005	16
530630024002030	Block 2030	9	530630016001011	Block 1011	14
530630024001027	Block 1027	0	530630024001103	Block 1103	12
530630024001044	Block 1044	8	530630016002012	Block 2012	0
530630023001006	Block 1006	40	530630016002019	Block 2019	10
530630023001005	Block 1005	24	530630016001037	Block 1037	12
530630104011022	Block 1022	71	530630024001050	Block 1050	0

Appendix E cont'd

GEOID10	NAME10	POP10	GEOID10	NAME10	POP10
530630024001097	Block 1097	0	530630016001054	Block 1054	0
530630024001005	Block 1005	37	530630111022008	Block 2008	37
530630024001001	Block 1001	3	530630020004036	Block 4036	26
530630024001011	Block 1011	1	530630024001075	Block 1075	59
530630024001000	Block 1000	0	530630024001058	Block 1058	0
530630024001007	Block 1007	0	530630111011008	Block 1008	0
530630020004026	Block 4026	10	530630020004017	Block 4017	16
530630023001008	Block 1008	31	530630020004025	Block 4025	25
530630016002009	Block 2009	37	530630111011007	Block 1007	0
530630016002022	Block 2022	15	530630024002010	Block 2010	15
530630024002009	Block 2009	25	530630024002016	Block 2016	0
530630024002001	Block 2001	99	530630024002025	Block 2025	26
530630024002017	Block 2017	36	530630024001032	Block 1032	0
530630024001030	Block 1030	19	530630024001095	Block 1095	0
530630020004035	Block 4035	29	530630024001045	Block 1045	0
530630024001073	Block 1073	3	530630024002028	Block 2028	19
530630024001112	Block 1112	2	530630024001024	Block 1024	0
530630035002053	Block 2053	20	530630026003011	Block 3011	4
530630002003037	Block 3037	0	530630024001035	Block 1035	0
530630024001021	Block 1021	0	530630026003047	Block 3047	6
530630020004012	Block 4012	20	530630023001034	Block 1034	23
530630020004002	Block 4002	15	530630035002022	Block 2022	0
530630020004009	Block 4009	25	530630023001018	Block 1018	11
530630023001000	Block 1000	0	530630023001011	Block 1011	20
530630024001072	Block 1072	10	530630023001023	Block 1023	20
530630020004059	Block 4059	0	530630023001032	Block 1032	7
530630024001113	Block 1113	7	530630023001001	Block 1001	0
530630024001121	Block 1121	0	530630023001022	Block 1022	60
530630009001004	Block 1004	43			

Appendix F: Missing Sidewalk Segments with Equity Importance

SW_ID	Shape Length (Feet)	Focus Census Block Pop Sum	All Census Block Pop Sum	Focus Pop per foot	All Pop per foot	Overall Importance $r=1$	Overall Importance $r=2$	Overall Importance $r=0.5$
3535	311.691	5	5	0.01604155	0.01604155	0.03208309	0.04812464	0.02406232
3547	56.998	12	12	0.21053252	0.21053252	0.42106503	0.63159755	0.31579878
3559	206.448	23	39	0.11140803	0.18890926	0.30031729	0.41172532	0.24461328
3834	283.480	15	15	0.05291370	0.05291370	0.10582740	0.15874110	0.07937055
3873	339.921	19	19	0.05589536	0.05589536	0.11179071	0.16768607	0.08384304
3895	324.759	4	4	0.01231683	0.01231683	0.02463366	0.03695049	0.01847524
4134	155.178	4	4	0.02577692	0.02577692	0.05155384	0.07733077	0.03866538
4135	150.473	186	186	1.23610144	1.23610144	2.47220288	3.70830432	1.85415216
4155	164.932	16	16	0.09700964	0.09700964	0.19401928	0.29102892	0.14551446
4213	331.425	7	7	0.02112089	0.02112089	0.04224178	0.06336268	0.03168134
4214	167.970	4	4	0.02381383	0.02381383	0.04762766	0.07144149	0.03572075
4345	159.208	186	186	1.16828341	1.16828341	2.33656683	3.50485024	1.75242512
4422	191.728	1	1	0.00521571	0.00521571	0.01043142	0.01564714	0.00782357
4672	341.956	15	15	0.04386531	0.04386531	0.08773063	0.13159594	0.06579797
4691	343.226	12	12	0.03496240	0.03496240	0.06992480	0.10488719	0.05244360
4745	141.861	86	86	0.60622754	0.60622754	1.21245507	1.81868261	0.90934130
4757	144.313	86	86	0.59592539	0.59592539	1.19185078	1.78777618	0.89388809
4802	145.468	29	29	0.19935723	0.19935723	0.39871447	0.59807170	0.29903585
4877	339.215	22	22	0.06485564	0.06485564	0.12971127	0.19456691	0.09728346
4889	148.430	148	148	0.99710387	0.99710387	1.99420775	2.99131162	1.49565581
4890	142.864	148	148	1.03595144	1.03595144	2.07190288	3.10785432	1.55392716
4998	121.417	29	29	0.23884617	0.23884617	0.47769235	0.71653852	0.35826926
5075	21.593	7	7	0.32417608	0.32417608	0.64835216	0.97252824	0.48626412

Appendix F cont'd

SW_ID	Shape Length (Feet)	Focus Census Block Pop Sum	All Census Block Pop Sum	Focus Pop per foot	All Pop per foot	Overall Importance r=1	Overall Importance r=2	Overall Importance r=0.5
5229	332.979	61	61	0.18319469	0.18319469	0.36638938	0.54958407	0.27479204
5230	36.660	61	61	1.66391945	1.66391945	3.32783891	4.99175836	2.49587918
8143	31.009	29	29	0.93520171	0.93520171	1.87040343	2.80560514	1.40280257
9379	108.794	140	140	1.28683986	1.28683986	2.57367972	3.86051958	1.93025979
9605	171.625	18	18	0.10487991	0.10487991	0.20975983	0.31463974	0.15731987
10501	436.771	23	23	0.05265916	0.05265916	0.10531832	0.15797747	0.07898874
10520	219.116	85	85	0.38792188	0.38792188	0.77584375	1.16376563	0.58188281
10583	298.856	16	35	0.05353748	0.11711325	0.17065073	0.22418821	0.14388199
10584	298.576	17	36	0.05693700	0.12057246	0.17750946	0.23444646	0.14904096
10587	18.224	15	15	0.82310359	0.82310359	1.64620718	2.46931077	1.23465538
10588	18.556	17	36	0.91612383	1.94002694	2.85615078	3.77227461	2.39808886
10607	600.660	23	23	0.03829121	0.03829121	0.07658241	0.11487362	0.05743681
10618	621.537	2	2	0.00321783	0.00321783	0.00643565	0.00965348	0.00482674
10619	609.677	124	124	0.20338643	0.20338643	0.40677285	0.61015928	0.30507964
11946	621.530	130	130	0.20916142	0.20916142	0.41832283	0.62748425	0.31374213
11947	238.757	148	148	0.61987642	0.61987642	1.23975285	1.85962927	0.92981464
13068	319.454	27	27	0.08451918	0.08451918	0.16903836	0.25355754	0.12677877
21035	12.785	11	11	0.86041284	0.86041284	1.72082568	2.58123851	1.29061926
21036	22.993	11	11	0.47841355	0.47841355	0.95682711	1.43524066	0.71762033
21071	48.621	11	11	0.22623883	0.22623883	0.45247766	0.67871649	0.33935825
21074	34.486	11	11	0.31896879	0.31896879	0.63793758	0.95690638	0.47845319
21075	12.101	11	11	0.90901228	0.90901228	1.81802456	2.72703684	1.36351842
21563	179.586	33	33	0.18375646	0.18375646	0.36751291	0.55126937	0.27563468
21571	62.978	11	11	0.17466427	0.17466427	0.34932854	0.52399281	0.26199640
21772	41.262	29	29	0.70282035	0.70282035	1.40564070	2.10846105	1.05423053
21775	48.622	33	33	0.67871147	0.67871147	1.35742294	2.03613441	1.01806721

APPENDIX

Appendix F cont'd

SW_ID	Shape Length (Feet)	Focus Census Block Pop Sum	All Census Block Pop Sum	Focus Pop per foot	All Pop per foot	Overall Importance $r=1$	Overall Importance $r=2$	Overall Importance $r=0.5$
22119	47.162	11	11	0.23323743	0.23323743	0.46647487	0.69971230	0.34985615
22120	9.536	33	33	3.46058946	3.46058946	6.92117892	10.38176838	5.19088419
22170	77.150	34	34	0.44070093	0.44070093	0.88140186	1.32210279	0.66105140
22315	134.446	29	29	0.21570048	0.21570048	0.43140095	0.64710143	0.32355071
22399	23.253	11	11	0.47306252	0.47306252	0.94612504	1.41918756	0.70959378
22403	9.972	33	33	3.30915318	3.30915318	6.61830636	9.92745954	4.96372977
22404	24.716	11	11	0.44506055	0.44506055	0.89012110	1.33518165	0.66759082
22418	16.302	11	11	0.67476210	0.67476210	1.34952419	2.02428629	1.01214314
22675	98.584	8	8	0.08114922	0.08114922	0.16229845	0.24344767	0.12172383
22676	23.444	8	8	0.34123793	0.34123793	0.68247587	1.02371380	0.51185690
22681	83.117	11	11	0.13234363	0.13234363	0.26468726	0.39703090	0.19851545
22682	65.146	11	11	0.16885173	0.16885173	0.33770346	0.50655520	0.25327760
22712	58.463	29	29	0.49604190	0.49604190	0.99208379	1.48812569	0.74406285
22827	166.450	16	16	0.09612502	0.09612502	0.19225005	0.28837507	0.14418754
22841	169.675	130	130	0.76617111	0.76617111	1.53234221	2.29851332	1.14925666
22860	137.138	123	123	0.89690432	0.89690432	1.79380864	2.69071297	1.34535648
23165	19.273	18	18	0.93393380	0.93393380	1.86786760	2.80180141	1.40090070
23172	153.837	91	91	0.59153387	0.59153387	1.18306775	1.77460162	0.88730081
23548	16.751	18	18	1.07458906	1.07458906	2.14917813	3.22376719	1.61188359
30767	266.528	12	31	0.04502337	0.11631038	0.16133376	0.20635713	0.13882207
30768	267.044	23	23	0.08612804	0.08612804	0.17225607	0.25838411	0.12919205
30984	424.257	32	68	0.07542597	0.16028019	0.23570617	0.31113214	0.19799318
36279	11.628	186	186	15.99630143	15.99630143	31.99260287	47.98890430	23.99445215
41189	147.217	210	210	1.42646999	1.42646999	2.85293997	4.27940996	2.13970498
42628	360.371	130	130	0.36073891	0.36073891	0.72147783	1.08221674	0.54110837
42765	195.685	1	1	0.00511025	0.00511025	0.01022050	0.01533075	0.00766538

Appendix G: Highest Priority Segments When $r = 1$

SW_ID	Population Sum	Population per foot	Shape Length (Feet)	Overall Importance $r=1$
36279	186	15.996301	11.63	31.992603
21140	155	29.746881	5.21	29.746881
13580	98	28.062986	3.49	28.062986
5331	331	23.972128	13.81	23.972128
37052	367	22.925570	16.01	22.925570
6728	329	21.152436	15.55	21.152436
21347	167	19.582913	8.53	19.582913
21820	121	19.291442	6.27	19.291442
21275	113	19.124957	5.91	19.124957
9098	171	18.984902	9.01	18.984902
21705	237	15.257223	15.53	15.257223
37802	232	14.085588	16.47	14.085588
30230	143	12.697787	11.26	12.697787
22698	161	12.470077	12.91	12.470077
6754	193	12.445421	15.51	12.445421
2991	191	12.348941	15.47	12.348941
32822	109	12.088804	9.02	12.088804
34389	232	12.068994	19.22	12.068994
37054	208	12.050828	17.26	12.050828
38273	217	11.931624	18.19	11.931624
6360	480	11.778891	40.75	11.778891
32359	199	11.200417	17.77	11.200417
21683	76	11.080866	6.86	11.080866
6759	193	11.012024	17.53	11.012024
38042	189	10.951605	17.26	10.951605

APPENDIX

Appendix G cont'd

SW_ID	Population Sum	Population per foot	Shape Length (Feet)	Overall Importance $r=1$
33618	199	10.893739	18.27	10.893739
3001	191	10.793924	17.70	10.793924
5557	208	9.051653	22.98	9.051653
7722	217	8.992519	24.13	8.992519
32152	127	8.626477	14.72	8.626477
23529	100	8.151781	12.27	8.151781
39963	139	8.133489	17.09	8.133489
13695	50	8.116701	6.16	8.116701
7723	189	8.055181	23.46	8.055181
21990	62	7.976910	7.77	7.976910
5261	127	7.574005	16.77	7.574005
41369	139	7.347300	18.92	7.347300
7237	342	7.330397	46.66	7.330397
32823	109	7.024040	15.52	7.024040
38295	125	7.002407	17.85	7.002407
22511	175	7.000416	25.00	7.000416
SUM Popolation	7597		651.67	

Appendix H: Highest Priority Segments When r = 2

SW_ID	Population Sum	Population per foot	Shape Length (Feet)	Overall Importance r=2
36279	186	15.996301	11.63	47.988904
21140	155	29.746881	5.21	29.746881
13580	98	28.062986	3.49	28.062986
5331	331	23.972128	13.81	23.972128
37052	367	22.925570	16.01	22.925570
6728	329	21.152436	15.55	21.152436
21347	167	19.582913	8.53	19.582913
21820	121	19.291442	6.27	19.291442
21275	113	19.124957	5.91	19.124957
9098	171	18.984902	9.01	18.984902
21705	237	15.257223	15.53	15.257223
37802	232	14.085588	16.47	14.085588
30230	143	12.697787	11.26	12.697787
22698	161	12.470077	12.91	12.470077
6754	193	12.445421	15.51	12.445421
2991	191	12.348941	15.47	12.348941
32822	109	12.088804	9.02	12.088804
34389	232	12.068994	19.22	12.068994
37054	208	12.050828	17.26	12.050828
38273	217	11.931624	18.19	11.931624
6360	480	11.778891	40.75	11.778891
32359	199	11.200417	17.77	11.200417
21683	76	11.080866	6.86	11.080866
6759	193	11.012024	17.53	11.012024
38042	189	10.951605	17.26	10.951605

APPENDIX

Appendix H cont'd

SW_ID	Population Sum	Population per foot	Shape Length (Feet)	Overall Importance $r=2$
33618	199	10.893739	18.27	10.893739
3001	191	10.793924	17.70	10.793924
22120	33	3.460589	9.54	10.381768
22403	33	3.309153	9.97	9.927460
5557	208	9.051653	22.98	9.051653
7722	217	8.992519	24.13	8.992519
32152	127	8.626477	14.72	8.626477
23529	100	8.151781	12.27	8.151781
39963	139	8.133489	17.09	8.133489
13695	50	8.116701	6.16	8.116701
7723	189	8.055181	23.46	8.055181
21990	62	7.976910	7.77	7.976910
5261	127	7.574005	16.77	7.574005
41369	139	7.347300	18.92	7.347300
7237	342	7.330397	46.66	7.330397
32823	109	7.024040	15.52	7.024040
38295	125	7.002407	17.85	7.002407
22511	175	7.000416	25.00	7.000416
Sum Population	7663		671.18	

Appendix I: Highest Priority Segments When $r = 0.5$

SW_ID	Population Sum	Population per foot	Shape Length (Feet)	Overall Importance $r=0.5$
21140	155	29.746881	5.21	29.746881
13580	98	28.062986	3.49	28.062986
36279	186	15.996301	11.63	23.994452
5331	331	23.972128	13.81	23.972128
37052	367	22.925570	16.01	22.925570
6728	329	21.152436	15.55	21.152436
21347	167	19.582913	8.53	19.582913
21820	121	19.291442	6.27	19.291442
21275	113	19.124957	5.91	19.124957
9098	171	18.984902	9.01	18.984902
21705	237	15.257223	15.53	15.257223
37802	232	14.085588	16.47	14.085588
30230	143	12.697787	11.26	12.697787
22698	161	12.470077	12.91	12.470077
6754	193	12.445421	15.51	12.445421
2991	191	12.348941	15.47	12.348941
32822	109	12.088804	9.02	12.088804
34389	232	12.068994	19.22	12.068994
37054	208	12.050828	17.26	12.050828
38273	217	11.931624	18.19	11.931624
6360	480	11.778891	40.75	11.778891
32359	199	11.200417	17.77	11.200417
21683	76	11.080866	6.86	11.080866
6759	193	11.012024	17.53	11.012024
38042	189	10.951605	17.26	10.951605

APPENDIX

Appendix I cont'd

SW_ID	Population Sum	Population per foot	Shape Length (Feet)	Overall Importance $r=0.5$
33618	199	10.893739	18.27	10.893739
3001	191	10.793924	17.70	10.793924
5557	208	9.051653	22.98	9.051653
7722	217	8.992519	24.13	8.992519
32152	127	8.626477	14.72	8.626477
23529	100	8.151781	12.27	8.151781
39963	139	8.133489	17.09	8.133489
13695	50	8.116701	6.16	8.116701
7723	189	8.055181	23.46	8.055181
21990	62	7.976910	7.77	7.976910
5261	127	7.574005	16.77	7.574005
41369	139	7.347300	18.92	7.347300
7237	342	7.330397	46.66	7.330397
32823	109	7.024040	15.52	7.024040
38295	125	7.002407	17.85	7.002407
22511	175	7.000416	25.00	7.000416
Sum Population	7597		651.67	

Appendix J: Highest Priority Segments When $r = 0$

SW_ID	Population Sum	Shape Length (Feet)	Overall Importance $r=0$
21140	155	5.21	29.746881
13580	98	3.49	28.062986
5331	331	13.81	23.972128
37052	367	16.01	22.925570
6728	329	15.55	21.152436
21347	167	8.53	19.582913
21820	121	6.27	19.291442
21275	113	5.91	19.124957
9098	171	9.01	18.984902
36279	186	11.63	15.996301
21705	237	15.53	15.257223
37802	232	16.47	14.085588
30230	143	11.26	12.697787
22698	161	12.91	12.470077
6754	193	15.51	12.445421
2991	191	15.47	12.348941
32822	109	9.02	12.088804
34389	232	19.22	12.068994
37054	208	17.26	12.050828
38273	217	18.19	11.931624
6360	480	40.75	11.778891
32359	199	17.77	11.200417
21683	76	6.86	11.080866
6759	193	17.53	11.012024
38042	189	17.26	10.951605
33618	199	18.27	10.893739

APPENDIX

Appendix J cont'd

SW_ID	Population Sum	Shape Length (Feet)	Overall Importance $r=0$
3001	191	17.70	10.793924
5557	208	22.98	9.051653
7722	217	24.13	8.992519
32152	127	14.72	8.626477
23529	100	12.27	8.151781
39963	139	17.09	8.133489
13695	50	6.16	8.116701
7723	189	23.46	8.055181
21990	62	7.77	7.976910
5261	127	16.77	7.574005
41369	139	18.92	7.347300
7237	342	46.66	7.330397
32823	109	15.52	7.024040
38295	125	17.85	7.002407
22511	175	25.00	7.000416
Sum Population	7597	651.67	